JUNGLE JIVE

SUSTAINING THE FORESTS OF SOUTHEAST ASIA

Published in 2016 by Connor Court Publishing Pty Ltd

Connor Court Publishing Pty Ltd
PO Box 7257
Redland Bay QLD 4165
sales@connorcourt.com
www.connorcourt.com
Phone 0497 900 685

ISBN: 978-1-925501-05-6

Cover design, Maria Giordano, picture "Jungle canopy", Shutterstock.

Printed in Australia

Distributed in Australia by Brumby Sunstate

World distribution: Ingram.

FOR THE JUNGLES OF SOUTHEAST ASIA, AND FOR HANNA, CAILIN AND EBEN

# TABLE OF CONTENTS

# PROLOGUE

## AN OPENING PROPOSITION

In contemplating this book I thought about an opening proposition. What do I want to say? What argument do I want to advance? What point of view do I want to put to you? What can we achieve together? Does this book have a cause – a passion if you like? Why am I writing it when I have heaps of other things I could be doing.

Well yes indeed, the book does have an important mission. Put simply it is that we need to better recognise the incredible values of Southeast Asia's tropical jungles, and figure out what we can do to help ensure that these values are safeguarded so the plants and animals that call these jungles home are sustained for their own and for our sake.

Young Sumatran orangutan: home is the jungles of Southeast Asia

If you are taking the trouble to read this book – and thank you – you are probably aware that the story about tropical jungles being the 'lungs of the earth' has been told repeatedly. Containing, as they do, about 80 per cent of the plant and animal species on the planet, yet occupying only about five percent of the land, tropical jungles are indeed extremely precious biological treasure houses.

Tales of looming disaster if the rate jungle destruction and clearance is not slowed have been well documented; as have the stories of imminent peril for these the richest ecosystems on the planet. A fair question to ask is has the recognition of these values and the threats they face had any appreciable impact on slowing the rate of clearance and degradation? We will examine this proposition in the pages ahead.

Generally the environmental movement has been at the forefront of fashioning and delivering the 'save-the-tropical-forests' message, and academic institutions and research campuses have heeded this call-to-arms. However, despite the passion and argument the tragedy of the tropical jungles continues apace.

Why is this when the case appears so compelling? Simple I think – the argument to save the jungles has been driven by emotion and a dictatorial dogma that countries, especially those hosting jungle tracts, have an implicit duty to protect them. Clearly this argument has failed to win the day. Jungle destruction has continued to be motivated by the reality that in economic terms jungles have little value and therefore should be converted to some other use capable of earning a dollar.

The reality is that jungle destruction has and continues to be driven by grinding poverty in the developing world and the legitimate wish of these countries to improve their economic and welfare standards. This situation is compounded by Western nations' avaricious consumption.

*Malaysian jungle: little value in economic terms*

Through necessity the tens of millions living within or close to jungles often think and live short term. For them life is about tomorrow's meal, getting the family through the week and acquiring life's basic needs. Such communities see the jungle as a 'local bank' with assets that can be converted into cash to support family and community survival. The motivating factor behind all this has and continues to be the hard reality that jungles have little inherent economic value and therefore need to be converted into something more commercially 'useful' or just gradually whittled away by the millions of rural poor.

*Palm oil plantation, Indonesia: converting jungle to commercially attractive crops*

Fundamentally there must be strong economic incentives to retain jungles rather than destroy them. Although not necessarily palatable to everyone, economic options that might help 'save-the-jungles' are becoming more apparent. These options offer a solution to jungle retention from the advance of agricultural development; conversion to other forms of land use, or piece meal destruction by local inhabitants.

It may seem a strange contradiction, but the alarming implications from the adverse impacts of climate change and the need to reduce – or offset – carbon dioxide emissions that contribute to the 'Greenhouse[1] Effect' might be part of the answer to jungle survival. So addressing climate change may be a significant part of the reason why the economics of jungle retention have the capacity to transform the imperative demanding that jungles be cleared to make way for something more 'useful'.

So in a world where money talks there now appears to be a basis for arguing that the long term protection and management of jungles could be based on an economic case absent from past debates. This case could achieve some positive climate change outcomes; yield wood based products, and save the lives of those astonishing jungle animals that share the planet with us.

Despite the overwhelming evidence that humanity has exceeded the 'limits-of-growth' of the planet, many nations continue to be driven by consumer frenzy. No matter that oil stocks are finite and climate change is real the thinking is that somehow lifestyles can be sustained by – in the case of energy – growing biofuels on jungle land to replace petrol. Amazingly this is frequently painted as a good environmental outcome with little thought given to the impact on the jungles themselves.

*Earth seen from Apollo space mission: strong evidence that the 'limits of growth' have been exceeded*

---

1 Greenhouse gases are a natural part of the atmosphere. They include water vapour, carbon dioxide, methane, nitrous oxide and some artificial chemicals, such as chlorofluorocarbons. Greenhouse gases absorb and reflect the sun's heat to maintain the Earth's surface temperature.

The concentration of water vapour, the most abundant greenhouse gas is highly variable. The concentrations of the other greenhouse gases are influenced by human activities, particularly burning fossil fuels (coal, oil and natural gas), agricultural activity and land clearing. Once released into the atmosphere greenhouse gases remain there for a long time increasing the concentrations of gases that trap heat.

Human activities have increased the level of carbon dioxide and other Greenhouse gases. This has increased the Greenhouse Effect, trapping more heat in the atmosphere and causing global temperatures to rise.

As we will see in the chapters ahead the economic 'drivers' behind the clearing of jungle for palm oil and other crops are responsible for millions of hectares of the most luxuriant plant growth on the planet being stripped away. With the removal of this jungle mantle animals like the Asian elephant, Sumatran tiger and orangutan are being pushed further down the road towards extinction. All this despite descriptions of jungles being precious, priceless and irreplaceable.

The ferment over the decades by an articulate and well resourced environmental movement may have elevated the prominence of the cause of saving tropical jungles, but it has done little to arrest their ongoing destruction. It seems clear to me that the solution to sustaining jungles must be based on accepted economics of both developed and developing economies that resonate on stock markets and in villages among indigenous communities and rural migrants.

In relation to perpetuating jungles, developing economic opportunities lie in part in the prospect of sustainable independently verified timber production and carbon credit trading. Both provide an opportunity to assign real dollar values to jungles and the capacity to disperse that income across a spectrum of interests – governments, companies with legal entitlements over land – right down to local communities. To put it in simple terms it is about establishing the case for developing compelling economic incentives to retain healthy, functioning jungles across Southeast Asia.

Implementing these economic opportunities offers the possibility of delivering what well intended environmental, academic and research advocates have been unable to achieve. That is, creating a set of circumstances where tropical jungles are seen as an economic asset, not a liability, and where governments, corporations and local villages have a vested interest in keeping trees standing. Protecting the array of animal inhabitants will be a collateral benefit, but a momentous achievement nonetheless.

In this context – while still having ground to cover – the management of tropical forests for timber production has advanced. Now more sophisticated practices and independent verification systems are aided by increasing value added timber product manufacturing. Along with carbon storage and trading this should lead to an increased recognition of the long term economic worth of jungles.

*Furniture from trees: helping establish the case for economic incentives to retain jungle*

*Log loading, Peninsula Malaysia: advances in management for timber production*

As you read on you will see that throughout human history trees have held a special place, not only as a source of food, shelter, medicines and fuel, but as a cultural cornerstone of many societies. Trees have often and are still viewed as spiritual cathedrals – places of reverence and homes of the gods. Today, when trees and jungles are recognised as a source of so many products essential to daily life, they remain special places of environmental, cultural and spiritual importance for many communities.

*Schwedagon Pagoda, Yangon, Myanmar: trees remain places of reverence and home of the gods*

There is no doubt that if humanity is to survive beyond the twenty first century – beyond the time when finite mineral and petrochemical supplies have been exhausted – trees will be a key ingredient of that survival. Trees hold the prospect of supplying renewable energy and sustainable, environmentally friendly products yet to be developed.

Anyway, enough for now, together we will explore the jungles of Southeast Asia in detail in the chapters ahead. We will also – hopefully and more adequately that I have done in this opening proposition – discuss the threats and potential solutions to our central goal of sustaining the renowned, amazing and economically valuable jungles of a very special part of our planet.

# 1

## CONSUMING THE EDIBLE SUN

### LEAVES – HOW THEY WORK

So here we are, but where to start to tell one of the great stories of our planet, the role and importance of the tropical jungles of Southeast Asia. Their diversity; the home they provide for countless animals large and small; their essential value to tens of millions of humans; the beauty and versatility of their timber, and other vital products they provide to sustain human health and existence.

We could start big by talking about the jungles and their global importance; their grandeur and extent; their many thousands of plants species – mighty trees and tiny herbs. We could enthuse about the resident animals – from elephants to ants and about emblematic species – like the tiger, rhinoceros and orangutan. But I think the best place to begin to explore these magnificent biological communities and to understand their central role in the mysteries of the web-of-life is to start small. No – not with the trees and other plants – although we will get to them, but with the fundamental building blocks of these resplendent celebrations of life – the leaves themselves.

To be clear at the outset – apart from a very few organisms that are able to draw energy from hot ocean vents – all life on the planet depends either directly or indirectly on energy captured from the sun. At the beginning of this chain are the plants, or more precisely their leaves that convert carbon dioxide and water into chemicals, like glucose that store the sun's energy.

Trees have been described as the biological miracle at the very core of human evolution, for not only do they provide the essential ingredients that have given rise to the creation of life, they hold the key to the continuity of humanity. At the most basic level, it is the leaves – in the green canopies – where the truly magic story of nature is to be found. Yes green is certainly the favourite colour of the plant world.

Within the green, thin and vein-covered fabric of leaves something truly miraculous happens that provides the spark of life. It is the intrigue of how leaves in a process called photosynthesis are able to turn carbon dioxide from the sky into giant trees and other plants. How within these green leaf factories the stuff we humans and other animals breathe out is combined with energy from the sun and water drawn up from the ground and turned into carbon based sugars and cellulose to make that renewable commodity wood. Yes it is truly nature's magic show – a Houdini act that is the basis for all life as we know it.

*Eucalyptus leaves: green leaf factories turning sunlight into carbon based sugars*

Photosynthesis literally means *light synthesis* – nature's way of capturing solar energy and storing carbon. Wood produced by photosynthesis and subsequent timber products store carbon throughout their lifetime – an important point we will return to later.

So photosynthesis – the ability of plants to produce carbon based sugars from sunlight, using carbon dioxide and water – takes place in specialised structures inside leaves called *chloroplasts*. Present in all green parts of plants, chloroplasts are where the work of capturing the sun's energy and converting it to sugars takes place.

Let us look a bit further at the process of photosynthesis. When rays of sunlight strike leaves, light wave lengths in the green spectrum bounce back towards our eyes – so we see green. The other light wave lengths – the reds, blues, indigoes and violets – are trapped in the chloroplasts where their energy is captured and turned into sugars. These simple sugars are combined into larger, more complex cellulose or lignin sugars, or stored as energy reserves in starches and tubers. It is these complex sugar compounds upon which the plants themselves, animals and indeed civilisations are built.

A description of photosynthesis that I like was written in 2008 by Tim Flannery.[2] He said:

> A lot of people somehow imagine trees grow from the ground, they don't, they grow from the air, they are congealed carbon dioxide and all of that carbon is stored in them, otherwise it would be out there in the atmosphere heating our planet.

A neat way of summarising photosynthesis don't you think?

Let us not get too caught up with figures, but a simple example from a neat little book: *The miracle of trees*[3] puts the process of photosynthesis into figures that provide a bit of real life context. The book explains that on a sunny day a 100 year old beech tree breathes in about 35,000 litres of air. From this air the tree extracts 10,000 litres or eighteen kilograms of carbon dioxide and produces twelve kilograms of sugars and thirteen kilograms of oxygen. The book asserts that walking in a forest on a sunny day you can literally feel the oxygen in the air. Try it and see what you think!

---

2 Tim Flannery is an Australian scientist, environmentalist and global warming activist. He was Australian of the Year in 2007; the Chief Commissioner of the Australian Climate Commission and held the Chair in Environmental Sustainability at Macquarie University in Sydney, Australia.
3 From Page 12 of: *The Miracle of Trees*, Olavi Huikari 2012 Walker Publishing, New York.

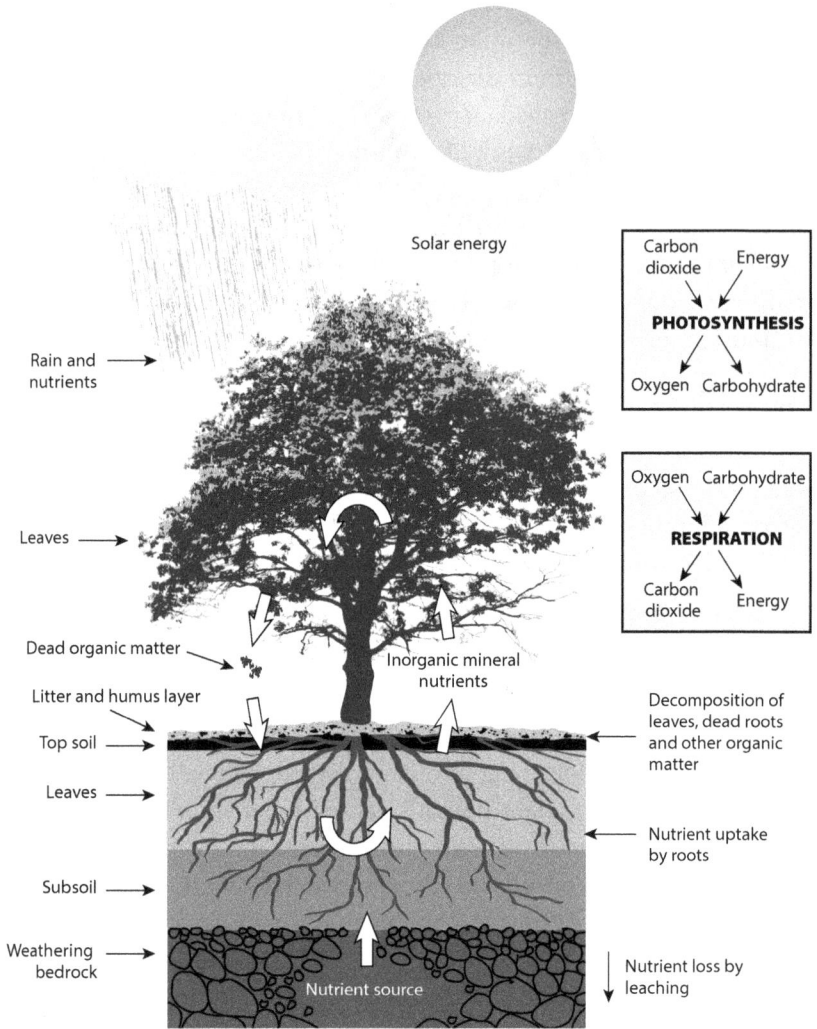

Solar energy

Rain and nutrients

Leaves

Dead organic matter

Litter and humus layer

Top soil

Leaves

Subsoil

Weathering bedrock

Inorganic mineral nutrients

Nutrient source

Carbon dioxide    Energy

**PHOTOSYNTHESIS**

Oxygen    Carbohydrate

Oxygen    Carbohydrate

**RESPIRATION**

Carbon dioxide    Energy

Decomposition of leaves, dead roots and other organic matter

Nutrient uptake by roots

Nutrient loss by leaching

*Photosynthesis and mineral cycles: The process of photosynthesis results in the production of oxygen and the absorption of carbon dioxide*

However, leaves and trees do a lot more than just produce oxygen and 'lock up' carbon dioxide. The world would be a much hotter place without them. Trees cool the air by shading and through water evaporation. They act like pumps to cycle water up from the ground back into the air. The 200,000 leaves on a large eucalyptus tree can take 40,000 litres of water from the soil and breathe it back into the air in a growing season. The cooling effect of all this water going into the atmosphere from just one tree is said to be equivalent to air conditioning twelve rooms.

So let us now look a bit more at this miracle of nature which is the reason why you and I, and all other animals exist – breath oxygen – and live off carbon based vegetation. Yes – the wood in our houses; the broccoli, beans and other vegetables on our plate; the meat we eat from animals turning green grass and plants into proteins and fats have all emerged from this mysterious piece of chemistry – photosynthesis. Without leaves capable of eating the sun we simply would not be here – plain and simple.

Stepping way back in time for a moment – when the Earth condensed out of a cloud of interstellar gas and dust, perhaps some four and a half billion years ago – the young planet was surrounded by a dense atmosphere of cosmic gases made up largely of carbon dioxide and carbon monoxide. As the planet began to cool and molten rock solidify, gases including water vapour and nitrogen were given off. However, this atmosphere was still a ghastly brew of toxic chemicals with little oxygen and not able to shield the Earth's surface from the scorching ultraviolet radiation from our young sun. Indeed, for the first primitive life forms that appeared about four billion years ago oxygen was deadly poisonous.

Then perhaps three billion years ago – give or take a few million – an evolutionary event of seminal importance occurred. Life forms capable of tolerating oxygen began to appear. Simple organisms in the sea started releasing oxygen that began to create a protective ozone shield and reservoir of free oxygen, allowing the first plants to emerge from the sea and occupy the land. Some millions of years later these simple life forms responsible for the oxygen rich atmosphere that made the evolution of animal life possible began to become more complex. And so the journey from simple plankton in the oceans to the complexities of tropical jungles continues to this day.

*Southeast Asian jungle: complex life forms from simple beginnings*

So life on planet Earth evolved within a thin, fragile envelope of soil, water and air. This, the biosphere, cocooned the planet's plant and animal life for hundreds of millions of years before our earliest human ancestors climbed out of the trees. Hop out into space and you will see this thin blue atmospheric band around the planet that stands between us and oblivion. The ongoing survival of life forms of every shape and intellect is absolutely dependent on the continuing efficient functioning of this biosphere. Unfortunately, as we will see, this is far from assured. Yes it was the trees – or more correctly their primitive plant ancestors – that actually created the conditions that made human and other forms of animal life possible.

*Planet Earth: a thin fragile envelope of soil, water and air*

Trees and humanity have always had a close working or *symbiotic* relationship. Animals consume leaves and breathe in the oxygen emitted by plants. In turn animals give off carbon dioxide that plants require in order to carry out photosynthesis and form carbon based sugars. Quite a neat circle don't you think?

The relationship between human beings, other forms of animal life and trees in their forest homes is described by Carl Sagan in his 1980 book *Cosmos.*[4] He says:

> Human beings grew up in forests; we have a natural affinity for them. How lovely a tree is, straining toward the sky. Trees are great and beautiful machines, powered by sunlight, taking in water from the ground and carbon dioxide from the air, converting these materials into food for their use and ours. ... And we animals, who are ultimately parasites on the

4 *Cosmos,* 1980, Random House, New York is a popular science book by astronomer and Pulitzer Prize-winning author Carl Sagan. Its thirteen illustrated chapters correspond to the thirteen episodes of the *Cosmos* TV series on which the book is based that explores the mutual development of science and human civilisation.

plants, steal the carbohydrates so we can go about our business. In eating plants we combine the carbohydrates with oxygen dissolved in our blood because of our penchant for breathing air, and so extract the energy that makes us go. In the process we exhale carbon dioxide, which the plants then recycle to make more carbohydrates. What a marvellous co-operative arrangement – plants and animals each inhaling the other's exhalations, the kind of mutual mouth-to-stomata resuscitation, the entire elegant cycle powered by a star a hundred and fifty million kilometres away.

Exciting isn't it, but you probably think this is all getting a bit too technical. Don't worry, all you really need to remember is that leaves, trees and other plants perform nature's truly magic show. Yes this conjuring act of nature has given birth to wonderful tropical jungles and their amazing biological cargo of hundreds of thousands of species of animals all evolving and continuing to live because of leaves. Indeed the green leaves are the fundamental wonder of life on Earth – you should never think of leaves in the same way again!

Let us now have a close look at plants themselves. Basically their primary function is to hold leaves up to the sunlight. And of course – if you are a leaf – the closer to the sun you can get the better off and efficient you will be. It's a pressure cooker situation in the dense jungles of Southeast Asia with thousands of plants on every hectare all elbowing each other to get the best place in the sun. It is most often the trees – some with mighty and massive trunks – that are able to get their leaves up beyond other plants and capture the sunlight. So in the jungle being tall is good.

In Southeast Asia trees of all shapes and sizes occupy every gap and layer in the jungle capturing sunlight and turning it into those important carbon based sugars. Building their wooden trunks from carbon and getting their leaves up into the sky as far as possible the great towering trees of tropical jungles are nature's feat of engineering supremacy.

The battle among plants for room and sunlight has led to efficiency improvements in their leaves. Leaves with more veins can carry more water allowing them to make more sugars ensuring that plants grow faster. In turn these plants can hold their leaves aloft to occupy more space and consume more sunlight. Through time plants that have been able to produce more and more veins in their leaves have won a greater number of races over the millions of years that the plant Olympics have been held.

*Malaysian jungle scene: thousands of plants elbowing each other to get the best place in the sun*

The character and behaviour of leaves divides them into two distinct groupings. Trees that have green leaves for more than one season are called *evergreens*. Those that lose their leaves due to seasonal climate changes are called *deciduous*. So there are evergreen trees and deciduous trees. In cooler, temperate climates deciduous trees begin losing their leaves in the autumn. As winter advances all the leaves are lost and trees stop growing until the warmer weather returns in spring when a new cloak of leaves appears.

In tropical climates with distinct wet and dry seasons, some trees lose leaves at the end of the wet season and remain dormant through the dry, but more usually tropical evergreen trees – as the name suggests – retain their leaves all year round.

*Mature eucalypt tree: evergreen trees retain their leaves all year around*

As deciduous trees prepare for their annual leaf loss they absorb the chlorophyll, plus other nutrients from their leaves. Without the chlorophyll pigment that colours the leaves green other pigments dominate and the leaves turn to shades of yellow, red, orange and brown. Once all the chlorophyll has been absorbed the leaves wither and fall to the ground. If you live in a cooler, temperate climate I am sure you would have raked up leaves during the winter to keep the yard tidy – they make a great addition to the compost bin.

Despite their common function and structure the size, shape and arrangement of leaves varies tremendously – large, small, thick, thin, compound, simple, curved or lobed. Leaf characteristics are largely determined by the climate in which a particular tree grows, but also by competition and 'attack' by leaf eating animals and insects.

Leaves are composed of an outer waxy layer and *epidermis*[5] cells that protect the interior cells where the photosynthesis takes place. Various pigments also help protect cells from damaging ultraviolet solar radiation. Some trees alter their leaf angle to further reduce the impact of the sun. For example, the leaves on a number of Australian eucalyptus trees hang down to minimise the heat of the midday sun prompting some to suggest that such trees look half dead.

In the tropical climate of Southeast Asia, tree leaves are frequently robust and on the larger side in order to withstand the hot conditions and frequent heavy rain. Leaves also often have distinctive long tips that allow rain water to drain from their surface during frequent tropical downpours.

Other leaf characteristics seem to have much to do with the battles of evolution that have gone on between plants for more than 400 million years. As we have seen, plants fight for sunlight and they fight for nutrients and water. Competition is why trees grow tall, stems become trunks and forests grow dense. The highest leaves do best and so trees have evolved to be as tall as possible within the limits of structural physics and gravity, rainfall, temperature and altitude.

Another thing, if you take the time to look at a leaf under a microscope, you will see what are indeed tiny holes in the surface. These holes are called stomata. They open and close and it is through these openings that carbon dioxide is absorbed and oxygen and water vapour released.

If you think life is tough – try being a plant! In the hot house that is the Southeast Asian jungle plants have more to cope with than just competition among themselves. The evidence of animals and insects of every shape and size eating leaves is as ancient as the evidence of leaves themselves. Certainly it isn't easy being green. Plants cannot run and hide, so they and their leaves have had to resort to all sorts of self defence mechanisms. Some produce chemicals to make them unpalatable, or even poisonous. Others ooze latex or have an armament of prickles, spikes or stingy hairs to protect them from being eaten.

---

5 The outer single-layered group of cells covering a plant, especially their leaves. The epidermis serves several functions, including protection against water loss, regulation of the exchange of gases, and absorption of water and mineral nutrients.

*Microscope view of leaf underside stomata: open and close absorbing carbon dioxide and releasing oxygen and water vapour*

Geared for growth and reproduction jungle trees and other plants engage in a variety of tricks in their bid to reproduce. Many enlist the help of middlemen, such as bats, bees, beetles and birds – attracting them with their bright colour, sweet nectar, fleshy fruit or strong smell. These are specialised relationships where the insect or animal acts as the accidental pollinator and in turn is rewarded by the plant with a meal.

Ultimately we humans are all parasites on plants, and in particular on trees. Trees are the major contributors of the oxygen we need to breathe and survive. If we fail to protect them, we shall ultimately perish with them. Yes indeed trees are the magic green machines holding part of the promise – part of the solution – that will allow humanity to endure beyond current consumption patterns and lifestyles.

As we progress through this book we will talk about the wonder of Southeast Asian tropical jungles – their evolution and their biological complexity. But don't ever forget that it all starts with the often overlooked leaves and the magic trick of nature they perform turning sunlight, carbon dioxide and water into life – without them nothing else would happen.

# 2

# GETTING TO KNOW YOUR COUSINS
## TREES – NATURE'S ARCHITECTURAL AND ENGINEERING MARVELS

Look around you – most often trees are everywhere and include among their ranks the world's largest land based organisms. Trees are certainly one of nature's architectural and engineering marvels, yet to many, trees inhabit a mysterious world, a world we humans have always been connected to but know little about.

*Tane Mahuta, large kauri tree, Waipoua Forest, New Zealand: a tree can become huge*

In this chapter I will attempt to stretch your biological knowledge further by examining aspects of the tree world – the way they function and grow – plus their geographic spread and diversity. Pretty much anywhere on the Earth's surface where there is sunlight, warmth, nutrients in the soil, plus water plants, including trees grow.

Let us start by trying to define just what is a tree. Most, I think would agree that the basic features of a tree are that they are typically tall and long lived. Yes – trees are simply plants that have learnt to grow high and live for a long time. They grow tall to compete with other trees – racing upward and spreading outward for sunlight and water.

Also important, but less obvious, trees grow from the top. Some other plants that may also be tall, such as bananas and some palms grow from their base – that is from the ground. They are not trees. A tree usually has a rigid, woody, strong expanding trunk, or trunks, encased and protected by a layer of bark. These trunks support crowns of branches, twigs and leaves.

Trees also have a complex root system acting both to anchor the tree to the ground and allow water and nutrients to be absorbed. With the constant imperative of seeking resources from the sky and the soil, given sufficient time, a tree can become huge – they just keep on growing.

Trees come in an amazing variety of forms from tall and narrow – as with many conifers – to the broad and spreading form of European oaks or African umbrella thorn trees. A tree's height and shape is determined both by its genetic 'blueprint' and its environment.

*African savannah, Serengeti, Tanzania: height and shape determined by genetics and environment*

Not only do we need to define a tree, but also to classify them. There is a real risk here that we could rapidly get really complicated and move into the somewhat dark and mysterious world of taxonomy. Taxonomists – a group of scientists that continue the ancient tradition originating from Europe of giving all organisms scientific names based on Latin or Latinized Greek. As far as we can we are going to steer clear of such areas of scientific intrigue and difficult language, suffice to say that we know from the previous chapter that various broad classifications can be applied to trees and other plants. We have already learnt about evergreen and deciduous trees.

Biologists – and indeed those intriguing taxonomists – regard what is called *natural classification* as the useful way to order plants where plants are grouped together on the basis of common descent from an ancestor from back down the evolutionary pathway.

To keep it simple we are going to mention the two major natural classifications that are useful when it comes to trees – *gymnosperms* and *angiosperms*. At a fundamental level gymnosperms are the more primitive of the two categories and were the first to evolve. A key characteristic is that gymnosperms reproduce by means of an exposed seed – *gymnosperm* literally means naked seed.

From a tree identification perspective, conifers – the pines, firs and spruces – make up a very substantial proportion of the gymnosperms. Conifer means 'cone bearer' and refers to the way these trees carry their seeds on the scales of cones, rather than enclosed in a fruit developed from a flower, as is the case with the angiosperms also called flowering plants. Some conifers, like podocarp[6] trees bear their seeds on structures that are hardly recognisable as cones. Conifers also have pollen cones, smaller than the seed cones that produce wind borne pollen from numerous small pollen sacs, often seen as a distinct yellow cloud if you are fortunate enough to walk through a conifer forest when pollen is being released.

Although steadily pushed aside over geological time by the more advanced angiosperms, conifers include the world's tallest and longest lived trees and have remain a vigorous component of the world's plant community, especially in regions where the climate is cooler.

---

6 Podocarps are a large family of mainly Southern Hemisphere conifers, comprising about 156 species of evergreen trees and shrubs.

*European spruce: cone bearing conifers carry their seeds on the scales of cones*

*Giant redwoods, California, USA: confers include the world's tallest and longest lived trees*

From an evolutionary viewpoint the more recent angiosperms reproduce by means of a seed that is enclosed in an ovary of a flower. Angiosperms have colonised much of the world's land surface over the last 30 million years or so. They have evolved into the most diverse forms of plant life, with somewhere between 300,000 and 400,000 species. Many are grasses, orchards, daisies, legumes and so on, but in the tropics a substantial proportion of the angiosperms are the tall triumphant trees of the jungles.

To complete the picture – not to confuse you – angiosperms are also sometimes called broadleaved plants because they have 'normal' leaves as opposed to gymnosperms, like the pines and firs that have narrow needle-like leaves. It starts to get complicated, but hold on there is more. The terms hardwood and softwood are also frequently used when describing trees and wood so we should cover these terms well. They have a botanical basis, rather than referring to the hardness or softness of the wood.

*New Zealand pohutukawa flower blossom: a flowering broadleaf angiosperm*

The terms hardwood and softwood are synonymous with angiosperm and gymnosperm. Hardwood means the same as flowering, broadleaf angiosperm trees, whereas the softwoods are the gymnosperm conifers, firs and spruces. These terms can certainly give rise to some confusion. While the wood of most hardwood trees is in fact harder than that of most conifers there are some spectacular exceptions. For instance take balsa, a South American hardwood so light it was used by native inhabitants to build rafts[7]. More recently balsa has been used for model aeroplane and other light weight construction.

Another example is the Chinese princess tree or *Paulownia* – a light weight hardwood from China that for many centuries has been highly prized for furniture and other uses like cabinets, musical instruments and clogs. Today it also finds favour in aeronautical fit outs, surfboard cores and coffins.

At a microscopic level the cellular appearance of hardwood and softwood wood is quite different. We will get to wood properties later.

Going back to trees – really the term *tree* is somewhat arbitrary. Trees belong to many different families of plants and can vary drastically in appearance and growth habit. However, in the context of Southeast Asian jungles we are talking about fair dinkum trees – where tall trees dominate and characterise the landscape. Where the trunks and canopies of such trees are often home to other plants – epiphytes[8] and perching plants – together with many animals and insects.

You may have heard the well known poem by Joyce Kilmer[9] titled *Trees*. It is internationally famous and worth repeating here:

> *I think that I shall never see*
> *A poem lovely as a tree.*
> *A tree whose hungry mouth is prest*
> *Against the sweet earth's flowing breast;*
> *A tree that looks at God all day,*
> *And lifts her leafy arms to pray;*

---

7  The word balsa is Spanish for raft.
8  An epiphyte is a plant that grows harmlessly upon another plant, often a tree, and derives its moisture and nutrients from the air, rain, and sometimes from debris accumulating around it on the host plant.
9  Joyce Kilmer (1886-1918) was an American writer and poet mainly remembered for him poem *Trees* published in 1914. He was a prolific poet whose works celebrated the common beauty of the natural world and his religious faith.

*A tree that may in summer wear*
*A nest of robins in her hair;*
*Upon whose bosom snow has lain;*
*Who intimately lives with rain.*
*Poems are made by fools like me,*
*But only God can make a tree.*

Given what we now know about the evolutionary process, we can give Kilmer some licence in his reference to the origins of trees for such a great poem. However, Kilmer's sentiments were widely reflected throughout human existence. Trees have held a special place, not only as a source of food, shelter, medicines and fuel, but for numerous early societies as a cornerstone of their cultural and spiritual beliefs.

Trees have frequently been viewed as places of reverence and the home of gods. Even today when trees and forests are recognised as the source of so many products essential to daily life, they are still revered as special places of environmental, cultural and spiritual importance.

In his book; *The Tree in Changing Light*, Australian author Roger McDonald[10] expresses the connection between man and trees as follows:

It is impossible to separate trees from people's attitudes about themselves – their fears, their lack of self-acceptance, their timidity and their ignorance. But nothing is inflexible in human response. People can live and grow just as trees do, they can struggle and they can overcome what is in themselves.

Put another way, we humans share about half of our DNA[11] with trees – something for you to think about next time you sit under a tree!

A feature throughout human history has been the symbolism, ritualism and worship of trees. Across the spectrum of faiths, beliefs and cultures, trees have often marked the boundary between human understanding and divine mystery. Evidence for this goes back to the Stone Age more than 6000 years.

---

10 *The Tree in Changing Light*, Random House Books, Sydney 2001. Roger McDonald is a highly distinguished and prize winning Australian writer of eight novels and two works of nonfiction.
11 DNA or Deoxyribonucleic acid is a long molecule that encodes the genetic instructions used in the development and functioning of all living organisms. DNA is copied and inherited across generations. It is made of simple units that line up in a particular order within the DNA molecule. The order of these units carries genetic information, similar to how the order of letters on a page carries information. The language used by DNA is called the genetic code that provides the instructions for constructing and operating a living organism.

Humanity has had a deep, 'religious' relationship with trees. Yvonne Baskin[12] considers that humanity's oldest faiths and deepest symbols reflect an early, primitive connection to the natural world and with trees – imprinting on human consciousness a cyclic sense of death and decay, rebirth and renewal – so she said.

This relationship between trees and humanity is a dominant feature of religions. For example, the *tree of life* and *tree of the knowledge are* recurring themes in the Christian Bible. The Book of Genesis[13] states:

> And out of the ground the Lord God made every tree that is pleasant to the sight and good for food. The tree of life was also in the midst of the garden and the tree of the knowledge of good and evil.

There are frequent Biblical references to trees and gardens. The Book of Genesis also records humanity's first place of residence and sustenance as the Garden of Eden – one of the most indelible images of the Christian mythology. Elsewhere in the Bible's Old Testament there are stories of royal gardens, cultivations enclosed by walls and hedges, and vineyards where trees are a common element of the narrative.

Gods were sometimes thought to use trees as dwelling places, making such trees sacred. Sometimes gods spoke to early communities through the medium of a tree. In Greece, the oracle of Dodona[14] was given by the rustling of the leaves of the Oak of Zeus that was interpreted by officiating priestesses and priests.

In some early cultures, kings would sit beneath a majestic tree to administer justice. Sometimes the sacred tree constituted the heart of the city it protected. The Sumerians[15] worshipped Kiskanu, the cosmic tree in the

---

12 Yvonne Baskin is author of *The Work of Nature: How the diversity of life sustains us* (Island Press, 1997) and *A Plague of Rats and Rubbervines: The Growing Threat of Species Invasions* (Island Press, 2002). Her articles have appeared in *Science, Natural History, Discover,* and numerous other publications.

13 The Book of Genesis is the first book of the Hebrew Bible and of the Christian Bible's Old Testament. The basic narrative has as the central theme that God created the planet and all living things.

14 The shrine of Dodona was regarded as the oldest ancient Greece civilization oracle, possibly dating to the second millennium BC. Priestesses and priests in the sacred grove interpreted the rustling of the oak leaves to determine the correct actions to be taken. Aristotle considered the region around Dodona to have been part of Ancient Greece.

15 Sumerian is the common name given to the ancient non-Semitic inhabitants of southern Mesopotamia. They claimed that their civilization had been brought to the city of Eridu by their god Enki.

centre of Eridu, their holy city. On the Acropolis in Athens,[16] grew the olive tree planted by Athena, who had taken possession of the land and founded the city. In the Forum at Rome grew the fig tree beneath which Romulus and Remus[17] were suckled. According to Tacitus,[18] the withering of this tree in the year 58 AD was considered a bad omen.

According to Fred Hageneder[19] in his 2005 book *The Living Wisdom of Trees*, the entire spectrum of human existence is reflected in tree lore through the ages; from birth, death and rebirth to the age old struggle between good and evil, and the quest for beauty, truth and enlightenment. He writes:

> Whatever our personal beliefs regarding nature, spirits, and the question of whether God exists inside creation or only outside it (or at all), one thing is certain: the ability to extend compassion to other life forms, to feel gratitude and give thanks for sharing in the miracle of life, to respect, if not to love, all fellow inhabitants of this planet, makes us better human beings and helps us to triumph over ignorance and greed. The living wisdom of trees shows us that life is worth so much.

Hageneder suggests that the living wisdom of trees tells us that we are all travelling together through the cycle of life.

Try it if you haven't already there is something about standing next to a really big tree. Knowing that it is hundreds of years old – that it has been storing sugar and carbon in its trunk and pumping out oxygen for centuries – it somehow puts you and human life into perspective. Kind of makes you feel a bit puny and insignificant.

At the same time it is sad to think that it is humanity that has been responsible for wilfully destroying so many of these gentle, gracious defenceless giants that do nothing more than keep us alive.

---

16 The Acropolis of Athens is an ancient citadel located on a high rocky outcrop above the city of Athens and containing the remains of several ancient buildings of great architectural and historic significance, the most famous being the Parthenon.
17 Romulus and Remus were the twin brothers and central characters of the Rome foundation myth.
18 Publius Cornelius Tacitus was a senator and historian of the Roman Empire.
19 Fred Hageneder is a recognised authority on botany and trees. He has written several tree books and is a founding member and the chairman of Friends of the Trees, a registered charity concerned with nature conservation.

*Author John Halkett with large kauri tree: big trees put human life into perspective*

However, be aware – we will get to it later – I am certainly not saying we should not cut down and use trees – it is the how and the when that are the critical questions. Yes, while we do indeed recognise the biological wonder of trees, we should also see trees, forests and plantations in ways that work to produce the wide range of timber, paper and other products that are such an important part of our daily lives.

Cutting down and carefully utilising the right tree in the right place is a tribute to the value and respect society places on the qualities and attributes of trees. Indeed, the survival of humanity may depend on a better understanding and use of trees in ways that are only just being contemplated. As a means of mitigating climate change, as a source of energy, transport fuels, chemicals and other products yet to be invented.

*Tree felling, Malaysia: utilising the right tree in the right place*

Of course as we know, like people, trees live in families and communities. It is the mix of tree species that gives tree communities, or forests – or in the case of Southeast Asia jungles – their varied and distinctive characteristics. We will explore these forest homes in the next chapter.

# 3

## AT HOME WITH TREES

### FORESTS – PLACES WHERE TREES LIVE

Moving along – so far we have looked at nature's magicians – leaves turning sunshine into carbon based sugars by the process of photosynthesis – and at the trees themselves – feats of architectural and engineering excellence among nature's greatest achievements. So now let us turn to the places where trees – along with other plants, plus a host of animals dwell – the forests. In the case of this book particularly the jungles of Southeast Asia.

*African mahogany plantation: forests and plantations found in all regions capable of sustaining plant growth*

Forests are found in all regions of the Earth capable of sustaining plant growth. At altitudes up to the snow line, except where it is too dry, or where fire frequency is too high, or where the environment has been altered by natural disturbance or human activity.

Remarkably nature has gifted trees with the ability to survive in extreme weather conditions – from sub zero temperatures to dry scorching deserts. Trees grow in polar regions in Siberia, Alaska and Canada, where the ground can be frozen down to ten metres during periods of the year. By contrast in the dry Mojave desert of California the ten metre tall, 1000 year old Joshua[20] tree survives because of its spreading, deep penetrating roots reaching down to suck up any available moisture.

*Joshua tree, Mojave Desert, California: living on the edge of survival*

It is thought that forests cover about four billion hectares – about 30 per cent of the world's land surface.[21] However, the importance of forests

---

20 The Joshua tree (*Yacca brevifolia*) is native to southwestern North America in the states of California, Arizona, Utah and Nevada where it is confined mostly to the Mojave Desert. The Joshua Tree National Park located in southeastern California is named after the tree.
21 *State of the World's Forests 2012*, Food and Agriculture Organization of the United Nations, Rome 2012.

stretches far beyond their boundaries, because forests help to regulate the planet's climate. For example, according to the United Nations Food and Agricultural Organization they store nearly 300 billion tonnes of carbon in their living parts – roughly 40 times the annual Greenhouse gas emissions from fossil fuels.

It is estimated that around 13 million hectares of forests a year have been converted to other uses or lost through natural causes between 2000 and 2013 – a scary figure. However, annual tree plantation or forest rehabilitation is estimated at five million hectares. We will be returning to these figures later – they underpin much of the motivation for writing this book.

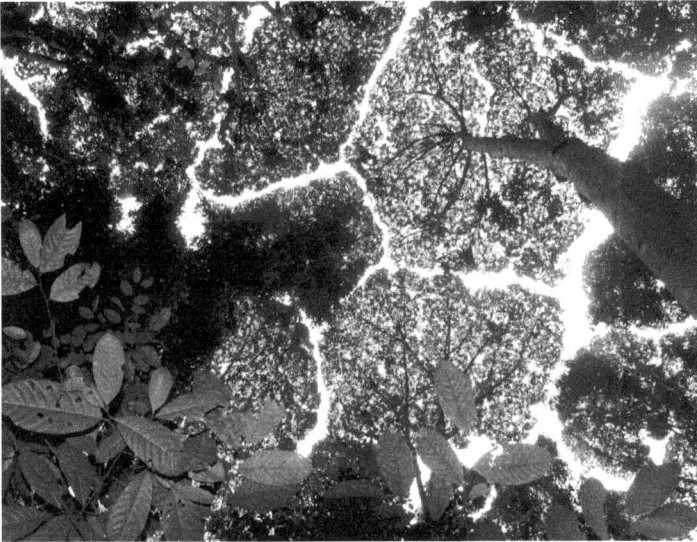

*Jungle canopy: loss of forest cover a scary statistic*

It is really the presence of trees that define – in descending order of tree cover or tree canopy density – forests, woodlands, shrublands, savannas and grasslands. However, as we have already acknowledged, there are large regions of the world where the march of civilization – settlements, gardens, fields and animals – has over the past couple of thousand years cleared large tracts of forest and other woody plant communities.

A few broad comments at this point might be useful. As a general rule, forests dominated by angiosperms are more species rich than those dominated by gymnosperms, although as always when it comes to plants there are some exceptions. Also for forests, biomass – the total amount of plant matter per unit area – is high compared to other vegetation communities.

In a forest branches and foliage of individual trees often touch or interlock, although gaps of varying size can exist. Woodlands have a more continuously open canopy, with trees spaced further apart.

Without getting too abstract, forests are one of the planet's great ecosystems – communities of living organisms and ecological systems interacting with the environment. Like other types of ecosystems forests deliver life's essential ingredients – clean air, pure water, habitat and resources. Plant and animal species are critical components – the cogs and wheels of functioning ecosystems. Lose too many species from a forest ecosystem and the whole system runs the risk of ceasing to operate effectively.

Forest cover varies enormously from one part of the world to another. It is largely climate[22] that determines height, complexity, tree density and species composition. As forest or woodland types change so too does the whole assemblage of resident animals and insects.

*Winter scene, boreal forest: climate determines tree height, tree density and species composition*

22 Notable among climatic variables that determine forest presence and characteristics are temperature, including both summer and winter averages and extremes, rainfall and its distribution over the seasons, humidity and wind.

Climate and soil properties greatly influence the distribution of forest and other tree-covered landscapes. Tall forest gives way to woodland where annual rainfall generally dips below about 700 to 800 millimetres although in some places a broad overlap zone exists.

As we have already seen for trees forest vegetation can be classified several different ways. An important attribute is structure, including things like tree density, composition and height; the presence or absence of understory trees and shrubs, and the presence of tree related growth forms, such as lianes, epiphytes or parasitic plants.

So at the risk of boring you, let us quickly look at some of the broad forest classifications. In terms of these classifications, again the approach in this chapter will be to keep things simple as there are probably as many ways to group, sort and sift forests as is the case for trees themselves.

We are going to mention just the very broad forest categories or types based primarily on latitude, although forests can also be classified based on the dominant tree species, resulting in numerous different forest types, such as ponderosa pine forest or beech forest. So for no particular reason starting in the north of the Northern Hemisphere we have the austere yet spectacular boreal[23] forests.

*Black bear: austere yet spectacular far northern forests*

---

23 The word 'boreal' means northern. Boreal forests are also known by the Russian name *taiga*.

For our purposes boreal forest occupies the northern sub arctic zone up beyond about latitude 50 and is generally comprised of evergreen conifer tree species. The boreal region encircles the earth at the top of the Northern Hemisphere across Russia, Scandinavia, Alaska and Canada. The boreal forest belt represents the world's largest land based ecosystem – a swath of confers, with some deciduous trees that act as part of the largest source and filter of freshwater on the planet. Beyond the northern limit of boreal forest lies bleak treeless arctic tundra and ice.

Temperatures in these boreal forests are usually extremely low with long winter seasons. The soil freezes – only thawing for a few months in the farthest northern forests. Winter air temperatures can fall to as low as minus 60 degrees Celsius. Soils are generally thin and nutrient poor. Most of the water is delivered in the form of snow – 40 to 100 centimetres a year.

As indicated, the main tree species in the northern boreal forests are conifers – pines, spruces, firs and larches adapted to very cold climatic conditions. Their needle like 'leaves' have a small surface area that, along with an increasing concentration of sugars and starches, plus specialized proteins in their sap, acts as a natural 'antifreeze'. In addition oils and resins present in leaves, bark and wood also act as a protective mechanism. Dormant buds at growing tips are each coated with resin to insulate them against the cold. Some deciduous trees, such as birches, alders, poplars and willows are also present in boreal forests.

There are no boreal forests at similar latitudes in the Southern Hemisphere as this region of the planet is occupied by ocean, ice and penguins!

Boreal forests support the world's largest caribou herd; the second-largest wolf population, and polar, black and grizzly bears. A number of other notable animal species also inhabit these climatically challenging forests, including moose, lynxes, foxes, deer, bats, woodpeckers and others able to tolerate the cold, harsh weather conditions.

*Bull moose resting in dense spruce forest: boreal forests support the world's largest caribou herd; the second largest wolf population, polar and grizzly bears, plus many other animals, including moose*

Temperate forests are located midway between the tropics and the poles. The climate is neither extremely hot nor extremely cold. The summers can feel hot and dry, but in these forests climate is never so harsh that the soil dries up or the plants die. Likewise, the winters may produce a lot of snow, but are not as severe as in boreal forests. Temperate forests are said to occupy the earth's Goldilocks[24] zone between the chilly northern boreal forests and the tropical jungles of the equatorial zone.

---

24 The young girl Goldilocks is a central character in a famous fairy tale *Goldilocks and the Three Bears* written in 1837 by British author and poet Robert Southey. Later adaptations of the original narrative tell the story of three bears – little bear, middle sized bear and big bear - who lived together in a house in the forest. One day they take a walk in the forest while their porridge cools. Goldilocks enters the house while the bears are out and tries the three bowls of porridge. She finds the first bowl too hot, the second bowl too cold, but the third bowl 'just right'. In colloquial use the expression *the Goldilocks zone* is used to suggest that the temperature or the environmental conditions are 'just right' - not too hot and not too cold.

Temperate forests are found in both hemispheres from latitudes 25 to 50 in regions of north eastern Asia, North America, western and central Europe, southern South America and Australasia. They can be coniferous, deciduous or contain mixed species depending on geography and climate.

Warm temperate zones also support broadleaf evergreen forests. An interesting thing about these forests is that they have all four seasons; summer, spring, winter and autumn, plus soils that are generally rich and fertile. These broadleaf forests include deciduous species like oak, maple, beech, hemlock, cotton wood, elm and many more.

Temperate conifer dominated forests include some of the world's tallest trees, such as those along the northwest coast of North America with the giant redwoods, Douglas fir and Sitka spruce. Other major areas of temperate conifer forest include the Northern Hemisphere forests of the mountains of western China, northeastern China and adjacent regions of Russia. Also in Japan, the mountains of central Asia and in the Himalayas conifer forests are present. Mexico's Sierra Madre ranges, central Europe the Balkans and Turkey all possess conifer dominated forests on their mountains as do the Atlas ranges of northwest Africa.

In the Southern Hemisphere there are smaller regions of conifer forests, such as the monkey puzzle tree, or *Araucaria* and the tall, long lived *Fitzroya cupressoides* conifer native to the Andes mountains of southern Chile and Argentina; the kauri and podocarp forests of New Zealand and some small patches of ancient conifers in Tasmania, Australia.

Vast tracts of former temperate deciduous dominated forest have been cleared for settlement and farming. Frequently derived from glacial deposits the underlying young soils are mostly fertile and moisture retentive. In large parts of China this clearance has been almost total; less so in Europe. In the United States of America settlers have cleared vast tracts of deciduous forest for cities and farms over the past four centuries.

In the Southern Hemisphere, broadleaf temperate forest includes the denser forests in cooler regions of South America – really only in Chile and Argentina, in Australia, New Zealand and the southern tip of Africa. In Australia eucalypts account for more than 70 per cent of trees in forests and woodlands, growing in a wide range of climates from the hot tropics to near desert inland plains and up on to alpine snow fields.

*Northern temperate deciduous forest: vast tracts cleared for settlement and farming*

Almost all eucalypts occur naturally only in Australia. This iconic plant group is comprised of more than 700 species of trees and shrubs. Evolving from rainforest ancestors, eucalypts have adapted to an environment where nutrient poor soils are common and a dry environment has become the norm.

Many eucalypts secrete a resinous gum – hence the name by which they are known around the world – gum trees. They have distinctive foliage, multi-stamened flowers and seeds that are contained in woody and sometimes hard capsules, protecting them from fires and insects.

Ashley Hay[25] says of the gum tree:

> Over 700 species make up the genus and, with a dozen exceptions, this is the only place where they grow naturally - from shaggy-barked blackbutts on the east coast's dunes, to the bright bark of snow gums up at the alpine tree line, to silky-white ghost gums in the centre. There are scrubby mallees that won't pass shoulder-height, and pillars of mountain ash that shoot a hundred metres into the sky: in Western Australia there are dark-leaved jarrahs that can grow to either of these heights. Site-specific, they carve the country into distinct ribbons and patches: one species grows up a hill and stops where another suddenly starts and grows down the other side.

---

25 In her outstanding book simply titled: *Gum* 2002, Duffy & Snellgrove, Potts Point, NSW.

*Tasmanian eucalypt forest: more than 700 species known around the world as gum trees*

Some areas of Australia's island state of Tasmania are dominated by myrtle forest (*Nothofagus*) with strong similarities to the extensive beech forests (also *Nothofagus*) in the South Island of New Zealand.

The latitudes about ten degrees either side of the Equator are mostly covered in tropical jungle or rainforest that are of particular interest in this book. Describing tropical forests, but equally applicable to other forest systems, in her 2002 book *Green Malaysia – Rainforest Encounters*[26] Premilla Mohanlall usefully summarised the forest as follows:

> All forests are dynamic environments which, once established, are hardy, self-renewing machines. While nourishing themselves, they also share their bounties with other life forms, feeding and fertilising them in a manner befitting their status as Mother Nature.

We will get to tropical jungles in much more detail in the next chapter.

---

26 *Green Malaysia – Rainforest Encounters*, Premilla Mohanlall 2002, Malaysian Timber Council, Kuala Lumpur, Malaysia.

# 4

# CASTLES BUILT ON SAND

## JUNGLES – THE STORY BEGINS


So now we are going to get down to the essence of this book – tropical jungles – particularly those across the length and breadth of Southeast Asia. From early times tropical jungles in the world's hot and moist equatorial zone have captured the imagination of poets, writers, artists and explorers as dark, mysterious, brooding keepers of the planet's secrets. Their elaborate architecture contains extraordinarily rich and fragile habitats that house more than 50 per cent of the species living on the planet. Yes – the jungles are the Earth's green heritage with all of their beauty, diversity, wealth and fascination.

As a starting point to this chapter let us take a step back to some of the early written accounts of forests from what is generally known as the Age of Discovery.[27] Many of these science and geography narratives, including references to plants, animals and people of the New World[28] are attributed to early explorers.

Forests attracted the interest of navigators like Captain James Cook. Why him and not someone else – I was born in New Zealand and am writing this book in Australia and Cook commented on the forests in New Zealand in his journals and subsequently, when sailing along the eastern coast of Australia

27 The Age of Discovery was generally considered to be a period starting in the early 15th century and continuing to the 17th century. During this period European countries journeyed to and explored Africa, the Americas, Asia and the South Pacific.

The Portuguese began exploring the Atlantic coast of Africa from 1418. In 1492 Spain's Christopher Columbus sailed west to reach the Indies by crossing the Atlantic. However, he landed on an uncharted continent America then seen by Europeans as a new world.

After 1495, the French, English and the Dutch entered the exploration race venturing into the Pacific Ocean around South America, following the Portuguese around Africa into the Indian Ocean; discovering Australia in 1606, New Zealand in 1642, and Hawaii in 1778.

28 The New World was the name given to the Western Hemisphere, specifically the Americas. The name originated shortly after America was discovered by European explorers in the early 16th century, expanding the geographical horizon of the people of the Europe who up to that time had thought of the world consisted only of Europe, Asia and Africa, now sometimes collectively referred to as the Old World.

in the *Endeavour* in 1770, remarked on the apparently endless forests. He was quick to report back to the British Admiralty about the possibilities of timber for naval use.

*Captain James Cook: reported on forests along the eastern coast of Australia in 1770*

However, about 300 years before Cook sailed into southern oceans Christopher Columbus's[29] 1493 account of the island of Española[30] sets out what might be the earliest written description of the kind of vegetation with which this book is particularly concerned. In his journal of his first of three voyages[31] he wrote:

> ... filled with trees of a thousand kinds and tall, and they seem to touch the sky. And I am told that they never lose their foliage, as I understand it, for I saw them as green and as lovely as they are in Spain in May and some of them were flowering, some bearing fruit and some in another stage according to their nature.

29  Christopher Columbus (1451-1506) was a Spanish explorer, navigator, and colonizer completed four voyages across the Atlantic Ocean that led to general European awareness of the American continents.
     During his first voyage in 1492 Columbus landed in the Bahamas archipelago. Over the course of three more voyages, Columbus visited the Greater and Lesser Antilles, as well as the Caribbean coast of Venezuela and Central America, claiming them for the Spanish Empire.
     Columbus's voyages led to the first lasting European contact with the Americas, inaugurating a period of European exploration, conquest, and colonization that lasted for several centuries.
30  Española Island is part of the Galápagos Islands. The English named it *Hood Island* after Viscount Samuel Hood. It is located in the extreme southeast of the archipelago.
31  *Select Documents Illustrating the Four Voyages of Columbus*, Vol 2. (1930–33, reprinted 1967), Translated and edited by Cecil Jane, Kraus Reprint.

Tropical forest that in Columbus's time covered most of the West Indies[32] still stretch over vast tracts of the lowlands straddling the Equator in South and Central America, Africa and Southeast Asia. It is the characteristic vegetation of wet tropical areas with a hot humid climate and high, well distributed rainfall.

*Spanish explorer Christopher Columbus: earliest written description of jungle vegetation*

In botanical literature jungle is often referred to as tropical rainforest. In more popular use this rainforest is frequently just call *bush* or *jungle*. The name rainforest is not only given to the evergreen forests of the moist tropics, but is also applied to the somewhat less luxuriant and profuse evergreen forest found in subtropical and temperate climates in south western China, southern Chile, South Africa, New Zealand and parts of eastern Australia. In this book rainforest is used predominantly in the context of moist tropical forest or jungle.

I like to use the word *jungle* – it has a bit of a mysterious, romantic, exciting ring to it. Perhaps that comes from reading too much of Rudyard Kipling's[33] *The Jungle Book* when I was much younger. I do like Kipling's *Law of the Jungle* poem, the title of which is now in the common use as an expression suggesting the 'survival of the fittest' or similar attributes. Here it is:

---

32 The West Indies is a large group of islands that separate the Caribbean Sea from the Atlantic Ocean.
33 Rudyard Kipling (1865-1936) was an English short-story writer, poet, and novelist. He received the Nobel Prize in Literature in 1907. He was born in India, but taken by his family to England when he was five years old. Kipling was one of the most popular writers in England both in prose and verse in the late 19th and early 20th centuries. He is best known for his works of fiction, including *The Jungle Book*. This and other books are enduring classics of children's literature.

Now this is the Law of the Jungle – as old and as true as the sky;
And the Wolf that shall keep it may prosper, but the Wolf that shall break
it must die.

As the creeper that girdles the tree-trunk, the Law runneth forward
and back – For the strength of the Pack is the Wolf, and the strength of
the Wolf is the Pack.

Now these are the Laws of the Jungle, and many and mighty are they;
But the head and the hoof of the Law and the haunch and the hump
is – Obey!

Actually, like Kipling the word *jungle* has its origins in India. It is derived
from the Hindi word *jangal*, meaning impenetrable forest around settlements.

In the complex jungle system, the status quo is maintained by a variety
of relationships between life forms where collaboration, competition,
predation, resistance and specialisation are the order of the day.

A particular feature of jungle or tropical rainforest is that the overwhelming
majority of plants are trees of all shapes and sizes. Not only are trees the
dominant members of these communities, but most of the climbing plants
and some of the epiphytes are also woody. The undergrowth largely consists
of seedling and sapling trees, shrubs and young woody climbers.

*Tropical jungle: overwhelming majority of plants are trees of all shapes and sizes*

Can't image why, but I rather like this description of rainforest from my book *Trees that call Australia Home*[34] that refers to the tropical forests of the top eastern coast of Australia:

> Trees generally grow in shallow soils propping themselves up with the aid of distinctive buttress roots. Rainforests are also distinguished from other forest types by the presence of complex species composition, characteristic life forms, such as epiphytes, vines, tree ferns and palms. They usually have moisture loving, closely spaced trees with dense leafy canopies limiting the light penetrating to the ground. Often palms and tree seedlings poke up from the ground, while convoluted vines weave their way from one tree to another. Branches may be heavily weighed down by birds' nest ferns and orchids.

The trees of the tropical jungle are extremely numerous in terms of numbers of species and size, although the dimensions reached by tropical trees are sometimes exaggerated. The average height of the taller trees rarely exceeds 50 to 60 metres, though individual trees may grow up to 90 metres tall.

Buttress roots are a fascinating feature in tropical jungles. Unlike temperate forests where soils are frequently deep and rich in nutrients, jungle soils are generally thin and contain few nutrients. Because of this tree roots do not extend down very far into the soil, but instead spread outward just below the surface of the ground to absorb available nutrients. So because jungle trees are tall, they need extra support to keep them from toppling over. Buttress roots perform this function, growing like stilts from the base of trees, often very thick, they prop up trees giving them added strength and stability.

Though tropical jungle trees are sometimes taller than those in temperate forests they do not reach the gigantic dimensions of the Californian redwoods or the large eucalypt gums of Australia, both of which are recorded as exceeding heights of 100 metres. By way of further illustrating the contrast between forest types, in temperate forests dominant trees frequently belong to just a few and sometimes just a single species. By comparison in tropical rainforests there is seldom less that around 50 trees species present on any given hectare and frequently well over a 100. The richness and diversity of the tree flora is indeed an important characteristic of jungles and many of their other features depend on this richness and diversity.

---

34 *Trees that Call Australia Home*, 2008, Potts Points Publishing, Sydney, Australia.

*Young tree with buttress roots: provide extra support to prevent trees toppling over*

Tropical jungle forms a discontinuous band around the Earth, bisected somewhat unequally by the Equator, so that rather more jungle lies in the Northern than the Southern Hemisphere. The world's jungles occur in three big chunks. The largest continuous tract is found in South America, in the basin of the Amazon River. This massive area sometime called 'the lungs of the Earth' extends west to the lower slopes of the Andes mountain range and east to the Guianas[35] broken only by areas of savanna grassland and deciduous forest.

---

35 The Guianas refers to a region in north eastern South America which includes the three territories of French Guiana and Suriname. Sometimes the Guayana region, formerly the Guayana province, in southeast Venezuela; Portuguese Guiana, in northwest Brazil are also considered part of the Guianas.

Jungle extends south into the region of the Gran Chaco[36] and north along the eastern side of Central America to southern Mexico and to the chain of the Antilles Islands of the West Indies. In the extreme north west of South America, including in Ecuador and Columbia there is a narrow belt of jungle on the western side of the Andes Mountains. On the east coast of Brazil there is another narrow belt, separated from the Amazonian jungle expanse by a wide strip of deciduous forest.

Turning to Africa, the largest expanse of jungle lies in the Congo basin and extends westward into French Equatorial Africa, Gabon and Cameroon. Jungle continues as a narrow strip further west, parallel to the coast of the Gulf of Guinea through Nigeria and the Gold Coast to Liberia and French Guinea. In places the African jungle is fractured by regions of dry climate and open characteristic *Acacia* woodland and savanna grassland.

Bounded by China to the north, India to the west and Australasia to the east and south, the Southeast Asia region comprises a distinctive part of the Asian land mass. It is comprised of a seemingly innumerable collection of both large and small islands and is biologically amongst the richest and most complex areas on Earth. It is home to iconic animal species such as the tiger, orangutan, Asian elephant and rhinoceros – we will talk about them later – and also supports a number of animal species that have affiliations with Australia.

Southeast Asia has a predominantly equatorial, tropical climate and supports the world's third largest jungle tract after the Amazon and Congo. However, there is lots of variation, particularly where there are mountains. For example, Mount Kinabalu in Borneo – the highest point between the Himalayas and Puncak Jaya in Western Papua – adds to the region's plant biological diversity.

---

36 The Gran Chaco is a sparsely populated, hot and semi arid lowland natural region of the Río de la Plata basin, divided among eastern Bolivia, Paraguay, northern Argentina and a portion of the Brazilian states of Mato Grosso and Mato Grosso do Sul.

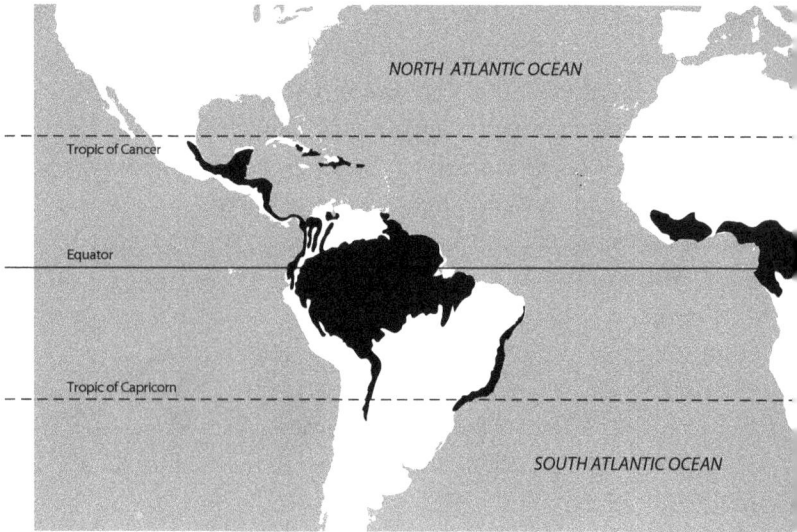

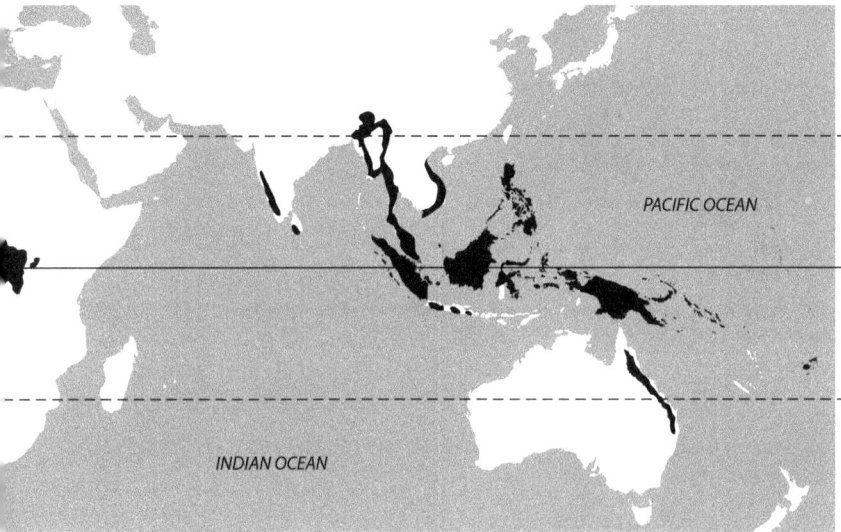

*Tropical forest distribution: The world's jungles occur in three big chunks forming a discontinuous band around the Earth, bisected somewhat unequally by the Equator*

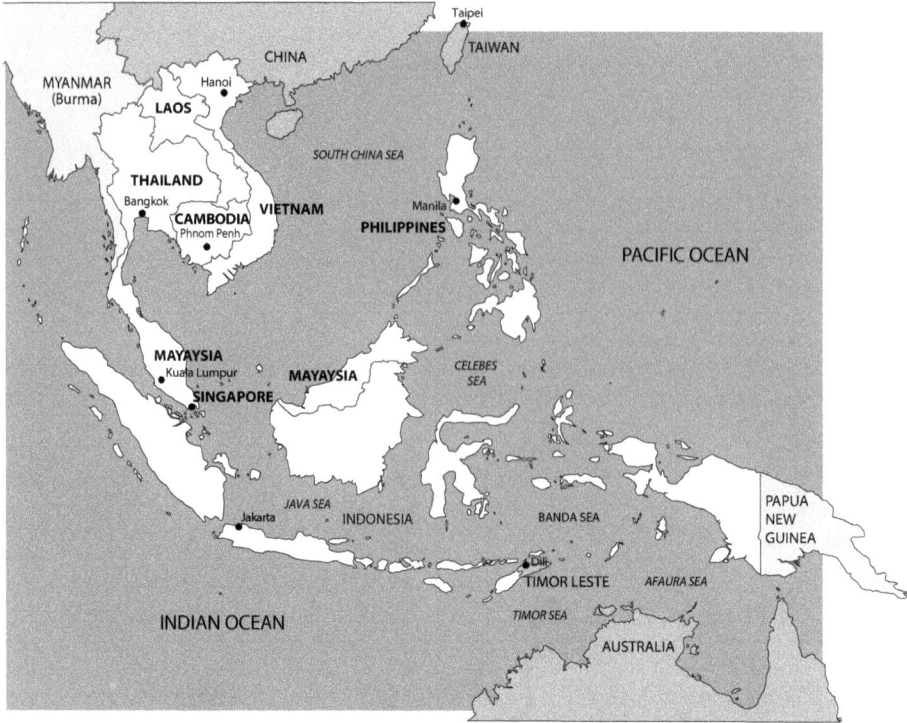

*Southeast Asia extent: purposes of this book Southeast Asia extends from Sri Lanka and western India to Myanmar and Thailand, then across Cambodia, Laos, Vietnam to the Philippines, as well as through the Malaysian archipelago, the islands that make up Indonesia and down to include New Guinea – Papua New Guinea and West Papua*

Let me be clear here, I am not going to be overly anxious in terms of geographic preciseness, so for the purposes of this book what I am saying is that Southeast Asia extends from Sri Lanka and western India to Myanmar and Thailand, then across Cambodia, Laos, Vietnam to the Philippines, as well as through the Malaysian archipelago, the islands large and small that make up Indonesia and down to include New Guinea – Papua New Guinea and West Papua. The largest relatively continuous piece of jungle in Southeast Asia comprises the island of Borneo – Sarawak and Kalimantan – together with Sumatra and New Guinea.

Down into Australia – not really Southeast Asia – tropical rainforest continues south as a narrow band along the eastern coast of the state of Queensland. Rainforest also extends out into the islands of the western

Pacific, including the Solomon Islands, Vanuatu, Fiji and Samoa. We will leave the description of Southeast Asian jungles here for the moment as we will be returning to them in more detail later.

Though at first tropical jungles may appear to be a bewildering chaos of vegetation they have a definite structure. However, in the often quoted phrase of Friedrich Junghuhn,[37] a jungle seems to show a *horror vacui* – from Latin meaning *a fear of empty space* – a term sometimes applied in forms of art where the entire canvas is filled with detail. In a jungle situation Junghuhn suggested that the forest seems anxious to fill every available space with trees, branches, twigs and leaves. However, a closer inspection shows that in the jungle trees form a number of structural elements of a relatively uniform vertical profile or pattern with a complex, but distinguishable spacial arrangement. With some variations these patterns are repeated throughout broadly recognisable tropical forest communities.

The fundamental building block – if you like – of tropical jungle is the soil. Perhaps the surprising reality is that apparently luxuriant, beautiful jungles are an illusion built on shaky, impoverished foundations. They are indeed *castles built on sand*, as the soils in many regions of the tropics are old, highly weathered, acidic and mineral poor. Certainly not the soil you would have in your garden if you wanted to grow cabbages or pumpkins. When jungle is cleared and the land sown in crops or pasture, soil fertility deteriorates rapidly.

Without tree cover the abundant rain and sun in the tropics badly damages soil. Planted crops or weeds cannot intercept and absorb the torrential rain that falls, so erosion and leaching depletes remaining nutrients. Without the constant shade of the jungle canopy, the soil gets a year round blistering from the hot tropical sun.

Putting this another way, because of the rapid decomposition of litter and the rapid recycling of nutrients back up into tree canopies tropical jungle soil remains nearly sterile. Given this reality governments, farmers and foresters should be

---

37 Friedrich Junghuhn (1809-1864) was an interesting German-Dutch botanist. As a student he was given to bouts of depression and he attempted suicide. He became involved in a 'matter of honour' and his opponent died of his wounds. Junghuhn fled by serving in the Prussian Army as a surgeon, but was discovered and sentenced to ten years in prison. He feigned insanity and was able to escape in the autumn of 1833. He was briefly a member of the French Foreign Legion in North Africa. He enlisted in the Dutch colonial army, arriving in Jakarta (then called Batavia) on October 1835 where he made an extensive study of the land and its people. He remained on Java until his death in 1864. The plants *Cyathea junghuhniana* and *Nepenthes junghuhnii* are named after him.

doing everything they can to protect these fragile tropical soils. Yet across the globe jungles are still being cleared wholesale at the rate of around 13 million hectares a year.

*Generalised tropical jungle vertical profile: Jungle trees form structural elements of a relatively uniform vertical profile or pattern with a complex, but distinguishable spacial arrangement. With some variations these patterns are repeated throughout broadly recognisable tropical forest communities*

However, the story gets worse – only about half of the jungle cleared actually expands the amount of land in agricultural production. The rest simply replaces the vast stretches of worn out land abandoned often after only a couple of years of agricultural use. Do the maths – it is not difficult – the rate of tropical deforestation could be halved without slowing the current growth of plantations, crops or grassland by just instituting modern agricultural practices, conserving soils and ending land abandonment.

Contrasted with temperate deciduous forests the reservoir of organic matter in tropical soils is small and short lived. Tribal people have always known this. In traditional slash-and-burn agriculture tribal jungle dwellers have, and still clear and crop small plots for only a few years, then move on leaving worked areas fallow, perhaps for a decade. If these fallow slash-and-burn areas have not been too heavily degraded, they are usually quickly recolonised by a vigorous succession of plants that restore soil fertility and productivity, and the diversity of the jungle.

So what then explains the deceptive lushness of tropical jungles? The answer is more about the rich and complex above ground plant communities than the attributes of the soil itself. It sounds – and indeed is a bit like a perpetual motion machine – where the jungle literally nourishes and replaces its own diversity, creating a far richer above ground plant 'paradise' than the tropical soil alone could ever sustain.

The reality is that nutrient levels in most tropical soils are so low that plant communities cannot survive if minerals are leached out of the soil by the frequent rain that drenches jungles. So when it comes to cycling nutrients jungle plant communities create an almost 'closed system'. Jungle trees establish a dense mat of fine roots just below the surface. This efficient root system 'sucks in' nutrients pretty much as soon as they are released from the litter.

The closed, multi layered jungle canopy captures a large proportion of rainfall. As rain runs across leaves and down tree trunks it picks up nutrients from the waste of insects and animals living in the canopy, along with substances leached from trees. Trunk water flow in central Amazonia is much richer in dissolved mineral nutrients than the original rainwater. Nitrogen compounds average fifteen times higher, phosphorus thirty times, potassium sixty times and calcium twenty-five times higher.[38]

With all its values and wonder the tropical jungle is regarded as nature's impressive supermarket and pharmacy. Almost half of today's main food crops were discovered in jungles, including bananas, mangoes, papayas, tea,

---

38 As detailed in: W J Funk and K Furch, 1985. *The physical and chemical properties of Amazonian waters and their relationships with the biota.* In G T Prance and T E Lovejoy (eds.) *Amazonia: Key Environments Series*, Pergamon Press, Oxford (1985).

rice, maize, various nuts, cloves, vanilla, cinnamon and coffee. Even the common domestic fowl originated from the jungles of India.

*Rooster and hens: domestic fowl originated from the jungles of India*

# 5

## On the road to Mandalay

### Myanmar – intriguing, diversity and disappearing jungles

Now we are going to turn directly to the jungles of Southeast Asia and add to the remarks in the last chapter about their amazing complexity and diversity. I am not going to attempt to track across every country in the region in a methodical way, but rather sample – if you like – enough jungle in different places to give you a flavour of its variation and geographic spread. Also to be honest, I will concentrate more on those countries with which I am more familiar, or where I have had more luck sourcing information that helps illustrate the arguments advanced in later chapters, or simply that I think you will find most interesting.

So from Myanmar in the north – down to New Guinea in the south, and of course the jungles of Indonesia – central to the Southeast Asia jungle story. As is Sarawak – the Malaysian portion of Borneo and Peninsula Malaysia. We may of course dip into other countries as well.

This means we will look at some countries as examples and also at notable features of the jungles across the region to give you a sense of things. The real purpose of this approach, as I have already said, is to get you up-to-speed so that we can traverse some of the critical policy issues, government actions and on-the-ground practices in later chapters. So let us see how we go.

Where to start – for no particular reason other than I have recently returned from another visit, I will say something about Burma. What do you know about Burma – or Myanmar as it is now more correctly called? The centre of Buddhism; historically part of the mystic, romance and intrigue of the old British Empire; and the focus of nostalgic poems by the Rudyard Kipling – yes him again! His 1892 poem *Mandalay* has been credited with compelling many to visit the country:

For the wind is in the palm-trees, and the Temple-bells they say:
"Come you back, you British soldier; come you back to Mandalay!"
Come you back to Mandalay.

Burma is also the early home of imperial British policeman turned author
George Orwell.[39] He spent his formative years there and his three famous
novels; *Burmese Days*, *Animal Farm* and *Nineteen Eighty-Four*, plus some of his other books are set in Myanmar or based on his experiences there.

To digress for a moment – because of Orwell's international standing as an author more than 60 years after his death, and his incisive observations whilst enforcing British law in Burma – it is interesting to read what he had to say about the harsh regime imposed on the country, and indeed other country 'possessions' during the days of the British Empire.

In his essay *Shooting an Elephant*[40] published in 1936, Orwell describes his distaste for and the brutality associated with the job he performed in the name of England:

> For at that time I had already made up my mind that imperialism was an evil thing and the sooner I chucked up my job and got out of it the better. Theoretically – and secretly, of course – I was all for the Burmese and all against their oppressors, the British. As for the job I was doing, I hated it more bitterly than I can perhaps make clear. In a job like that you see the dirty work of Empire at close quarters. The wretched prisoners huddling in the stinking cages of the lock-ups, the grey, cowed faces of the long-term convicts, the scarred buttocks of the men who had been flogged with bamboos – all these oppressed me with an intolerable sense of guilt.

*Rudyard Kipling and George Orwell: both feature prominently in the romantic history of Burma*

---

39 George Orwell was the pen name of Eric Arthur Blair, born in India in 1903 and died in London 1950.
40 *Shooting an Elephant and Other Essays*, George Orwell 2009 Penguin Books, London, UK.

I think this narrative is a powerful commentary on imperial British 'treatment' inflicted on many countries. But to return to Burma we also remember it as a fierce and bloody battle ground of the Second World War with its still haunting images of emaciated prisoners-of-war working and dying on the Japanese Thai-Burma railway.

We know that since independence from the British in 1948 Burma has been run by a military regime. One of the most closed counties on the planet with an appalling record of human rights abuses, ethnic genocide and descent into human misery and poverty. Its standing in the eyes of Western democracies has been abysmal. For nearly half a century, an army junta cowed the country and ruined its economy. Rape was used as a weapon of war in ethnic regions and children were enslaved. The military turned its guns on pro-democracy activists and protestors, most recently in 2007 when dozens of Buddhist monks were shot.[41]

But more recently, and suddenly, a transition towards democracy lead by Nobel Peace Prize recipient Aung San Suu Kyi has taken place. Truly a Nelson Mandela like figure after 15 years of house arrest she now strides the world stage helping to lead the country towards a brighter future and out of decades of debilitating international isolation and sanctions. Pleasingly, compared with the explosive revolutions of the so called Arab Spring, Myanmar's ongoing transformation has been much more peaceful.

Myanmar is now emerging as a geopolitical cornerstone wedged between the two most populous countries in the world, India and China. With its natural resources and positioned at a key global crossroad, Myanmar is also the newest economic frontier now that Western sanctions are being lifted.[42]

*Aung San Suu Kyi: set to lead Myanmar after 15 years of house arrest*

---

41 Further details are set out in the book: *Burma a nation at the crossroads*, 2012, Benedict Rogers. Random House, London.
42 Base in part on the article: *Inside Man* by Hannah Beech, *Time Magazine*, Pages 16-23. 21 January 2012 edition.

This background is by way of getting to the point that Myanmar still remains home to vast tracts of diverse tropical jungle, including about 90 per cent of the world's teak[43] trees. Myanmar's jungles are truly exciting, amazing places, still housing tigers and leopards. Elephants, rhinoceros, buffalo and boar are also resident, as well as many species of deer and antelope and an assortment of monkeys, flying foxes, wildcats and the tapir.

*Elephants at training camp: Myanmar still home to vast tracts of diverse jungle and animals*

Teak is usually the starting point in describing the jungles of Myanmar, often from a timber perspective as it is one of the world's most recognised and renowned timbers – exceptionally durable, stable, strong and attractive. In the 1930's well known architect Oliver Bernard[44] said:

---

43 Teak is the common name for the tropical hardwood tree species *Tectona grandis* and its timber products. It is native to south and Southeast Asia, mainly India, Myanmar, Indonesia, Malaysia and Thailand. It is also grown in plantations in many countries.

Teak is a large, deciduous tree that is dominant in mixed hardwood forests. It has small, fragrant white flowers and papery leaves that are often hairy on the lower surface. It is found in a variety of locations and climatic conditions from arid areas with only 500 mm of rain per year to very moist forests with up to 5,000 mm of rain a year. Typically, though, the annual rainfall in areas where teak grows averages 1,250-1,650 mm with a three to five month dry season.

44 Oliver Bernard (1881-1939) was an English architect and industrial designer.

The characteristics of teak are of infinite variety that if I were compelled
to reply on one timber for everything, teak would be my choice.[45]

There is evidence that the fundamental cause of the Anglo-Burmese
Wars[46] during the nineteenth century was conflict over the extraction of
Myanmar timber, particularly teak.

Myanmar's jungles vary from sub-alpine on the snow capped mountains
in the north, through dry and moist deciduous to tropical monsoon forests
in the south, to mangroves along the coast. There are thousands of recorded
plants, thousands of birds and hundreds of mammals and reptiles.

Myanmar is often mentioned as the last frontline of large scale
biodiversity. Biodiversity that is influenced by a wide range of factors such
as an latitude range between 9 and 28 degrees North; topography traversing
through three mountain ranges and four major river systems, and tropical
monsoon type climatic factors with three distinct seasons – the hot season
from mid February to mid May; the rainy season from mid May to mid
October and a cool season from mid October to mid February.


Sadly however, Myanmar's vital jungles are disappearing. The country has
one of the worst rates of deforestation on the planet, with 18 per cent of its
forests lost between 1990 and 2005.[47] Myanmar's forest administration has
been rife with corruption and illegality, leading to excessive logging and log
smuggling. Teak from Myanmar is especially sought after for its outstanding
characteristics. Since the late 1990s, neighbouring China has imported large
volumes of timber, much of which has been logged and traded illegally.

In 1997 China imported 300,000 cubic metres of timber from Myanmar;

45 From the *article A record price for quality natural teak* Ohn Gyaw, *Myanmar Forestry Journal* Vol 4
No 3 July, 2000, Yangon, Myanmar.
46 The First Anglo-Burmese War (March 1824 – February 1826) was the first of three wars
fought between the British and then Burmese Empire in the 19th century. The war, which began
primarily over the control of northeastern India, ended in a decisive British victory, giving the
British control of much of Burma. The Burmese were also forced to pay an indemnity of one
million pounds sterling and sign a commercial treaty.
   The war was the longest and most expensive war in British Indian history. Fifteen thousand
European and Indian soldiers died, together with an unknown number of Burmese army and
civilian casualties. The high cost of the campaign to the British led to a severe economic crisis in
British India in 1833.
   For the Burmese, it was the beginning of the end of their independence. The Burmese
would be crushed for years to come by repaying the large indemnity of one million pounds. The
British initiated two more wars against a much weakened Burma and assumed control of the
entire country in 1885.
47 UN- REDD Programme, Myanmar profile, 2012.

by 2005 this trade had risen to 1.6 million cubic metres.[48] Based on forest cover of 34 million hectares, or a bit over half of the total land area in 2005 and a current estimate of 1.4 per cent deforestation a year the forest area will decline to 27 million hectares or 41 per cent of the land area by 2020.

Myanmar's northern forests bordering the Chinese province of Yunnan have borne the brunt of the illegal logging onslaught. Biologically significant forests in Myanmar's mountainous Kachin State have been plundered to supply the Chinese market. Reports by the non government organisation Global Witness estimated that 900,000 cubic metres of illegal timber crossed the border from Myanmar to Yunnan in 2003, with two thirds of it coming from Kachin State. By 2005, the volume of logs from Myanmar imported by land into China's Yunnan Province reached one million cubic metres a year.[49]

In 2006 international exposure of the environmental and social costs of widespread illegal logging in northern Myanmar, plus conflict between insurgent groups and the military led to action being taken. The China and Myanmar governments signed agreements closing the border with Yunnan Province to timber trade. Instead log exports deemed legal by the Myanmar Government had to be shipped out from Yangon under the auspices of the state-run Myanmar Timber Enterprise. By 2008, illegal log imports across the Yunnan border had declined by 70 per cent.

Although large discrepancies in trade data exist between the China and Myanmar, the illicit log trade is thought to have risen again during recent years, reaching almost 500,000 cubic metres annually by mid 2012. This ongoing depletion of Myanmar's northern forests to supply the Chinese market is considered to be at crisis point. Timber traders in Kunming, the provincial capital of Yunnan, are now pushing ever deeper into Myanmar to secure supplies as areas closer to the border have been logged out.

Presently it appears that Chinese traders are suggesting that as long as taxes are paid at the point of import, logs are allowed into China despite a commitment from the Yunnan provincial government to allow only timber accompanied by documents from Myanmar authorities, attesting to its legal origin, to be exported into China.

The contrast in the condition of the forests along the border was striking;

---

48 Forest Trends, Baseline Study: Myanmar, 2011.
49 Global Witness, A Conflict of Interests, 2003.

while forests in the mountainous region on the Chinese side of the border are relatively intact, with large areas protected in the Gaoligong Nature Reserve, across the border in Myanmar's Kachin State the devastation wrought by excessive illegal logging is clearly visible. Chinese traders confirmed that supplies were also coming from further inside Kachin, as timber within a hundred kilometres of the border has already been logged. Traders also confirm that deals are done with insurgent groups to buy up areas for logging. One local community elder in Kachin interviewed by Environmental Investigation Agency[50] summed up the situation: "Myanmar is China's supermarket and Kachin State is their 7-11."

Across rural Myanmar it is apparent that for decades population pressure, poverty and lack of food security have contributed to serious jungle and environmental degradation. Presently about 70 per cent of the population live in rural areas and depend heavily on the jungles for their basic needs. Jungles continue to provide ethnic groups and people living in remote areas with posts, poles, fencing and household materials, fuel wood, fodder, and food, as well as wildlife for hunting.

In addition, inadequate electric power supply and limited provision of household gas fuel means a continuing reliance on wood and charcoal. It is estimated that consumption of fuel wood in Myanmar for 1990, 2000 and 2005 was 35.2, 40.6 and 44.6 million cubic metres respectively.[51] As for income generation, charcoal production is conducted for money making purposes in various regions of the country. Extracting non wood products, including wild fruit, latex, essential oils, wax, thatch, honey, bee wax, bat guano, orchids, edible birds' nests and medicines also support the meagre income of rural people.

The overall transformation of government in Myanmar to being much more liberal and transparent has been a remarkable culmination in the landslide election victory by Aung San Suu Kyi's pro-democracy party the National League for Democracy in November 2015. This achievement on top of the implementation of a log export ban April 2013 and reduction to annual log harvest volumes for teak and hardwood are significant and

50 *Appetite for Destruction - China's trade in Illegal timber*, Environmental Investigation Agency. 2012 London, UK.
51 *Myanmar forestry outlook study*, 2009 Asia-Pacific forestry sector outlook study 2, Working Paper No. APFSOS II/WP/2009/07, Khin Htun Food and Agriculture Organization, Bangkok.

promising signs for a much more sustainable future for the country's forest based industries.

The log export ban coupled with the reduction in the annual allowable log harvest has been accompanied by commendable efforts to comply with legality assurance requirements of leading consumer markets, and a willingness to develop a forest and timber products trade sustainability certification scheme. In addition, there has been a much increased focus on domestic value added processing, particularly of teak for high value marine decking and other applications. This effort is a pointer to the contribution the forest and wood processing industry is capable of making to Myanmar's longer term economic prosperity.

# 6

## DIPTEROCARP THE GREEN JEWEL

### MALAYSIA – GREEN HERITAGE AND DARK SECRET

Moving in a south easterly direction from Myanmar we encounter Malaysia. With a land area of about 33 million hectares Malaysia is a federation of thirteen states and three territories. Eleven of these states, plus the federal territories of Kuala Lumpur and Putrajaya covering 13 million hectares are situated on Peninsula Malaysia, while the states of Sabah, covering seven million hectares and the state of Sarawak, spanning more than 12 million hectares, are located across the Strait of Malacca on the island of Borneo.

Well forested, Malaysia has about 60 per cent of its land area still under forest – totalling about 19.5 million hectares.[52]

Arguably, Malaysia has a more sophisticated system of legally endorsed land tenure than some other Southeast Asian countries. Of the total area of forest about 75 per cent is set aside as Permanent Reserved Forests under Malaysia's 1985 *National Forestry Act* and related legislation. A further two million hectares have been declared national park or set aside as wildlife sanctuaries.

Malaysia's jungles are considered to be amongst the most species rich in the world, with many species found nowhere else. From the mighty dipterocarp[53] tree *keruing jarang*[54] to the largest pitcher plant[55] *Nepenthes rajah*, Malaysian

---

52 This comprises almost 6 million hectares in Peninsula Malaysia; 4.3 million hectares in Sabah and 9.2 million hectares in Sarawak.

53 The dipterocarps are members of the Dipterocarpaceae family of about 500 species of mainly tropical lowland forest trees. The largest genera are *Shorea* (196 species), *Hopea* (104 species), *Dipterocarpus* (70 species), and *Vatica* (65 species). Many are large emergent species, typically reach heights of 40-70 metres, some over 80 metres. Their distribution is from northern South America to Africa, the Seychelles, Sri Lanka, India, Indochina, Indonesia and Malaysia. The greatest diversity occurs in Borneo.

54 keruing jarang (*Dipterocarpus lamellatus*)

55 Pitcher plants are carnivorous plants with a prey trapping mechanism that features a deep cavity filled with liquid known as a pitfall trap. The Nepenthaceae and Sarraceniaceae families are the best known and largest groups of pitcher plants. The Nepenthaceae contains a single genus, *Nepenthes*, containing over 100 species and numerous hybrids and cultivars.

jungles occupy a multitude of habitats as a consequence of the geological history of the country, altitude range, the year round hot, moist climate and wide variety of soils. These factors determine the development, distribution and composition of the forests that cloak the country from the coast to the highlands.

At the risk of confusing you with figures, Malaysia's green heritage plays host to 14,500 species of flowering plants and trees, about a thousand species of animals and an estimated 20,000 to 80,000 species of insects – more than 6000 species of butterflies and moths – and an almost unfathomable number of other insects and life forms. Quite a bewildering inventory of life, don't you think? And all this density packed in the equatorial zone of uniformly high temperatures, high humidity and heavy rainfall – perfect conditions for luxuriant plant growth.

I love this description of Malaysian jungles from Premilla Mohanlall's 2002 book *Green Malaysia*[56] – *Rainforest Encounters* even though the words have wider relevance than just Malaysia:

> The rainforest is a dynamic environment, whirring away day and night like a lush green city that never sleeps. Its work is done quietly, often unnoticed, at all levels. Surfaces and strands of this mad muddle of towering, grasping clinging, strangling and lowly plants are tightly woven together in a complex and intricate web of habitats and ecosystems. Nowhere else is this interdependence illustrated than in the species-rich tropical rainforest, which teems with millions of life forms ranging from microscopic bacteria to fungi, fruit bats, birds, insects, lizards, tapirs and tigers acting out their respective roles as prey. Predator, pest pollinator and perpetuator of the species in the endless drama of life and death to maintain the stability of this closed system.

The dipterocarp family populates the multi layered tropical jungle that derives its name from its most common and famous inhabitant *keruing jarang*. The main difference between lowland and hill dipterocarp forests lies in their plant composition. A large number of species prefer warmer lowland temperatures, while some opt for the cooler highlands, with a few adapting to both environments.

Dipterocarp forests account for 85 per cent of Malaysia's jungles housing at least 270 out of the 502 known species of dipterocarps. The sheer size

---

56 *Green Malaysia – Rainforest Encounters*, Premilla Mohanlall 2002, Malaysian Timber Council, Kuala Lumpur, Malaysia.

and structure of these trees is awe inspiring. Sometimes referred to as the *Atlas*[57] tree of the jungle, their strong, tall trunks support other trees and plants. Rising up to 60 metres and spreading their crowns over a green domain as the emergent layer in this multi tiered celebration of life. The high and mighty dipterocarps distinguish themselves with their legendary 'roof garden on pillars' structure, not unlike a city of skyscrapers where everything races up to gain a place in the sun. Their contribution to jungle life is also to provide arboreal[58] animals with high rise homes from which they can move around without ever having to visit the ground.

Lowland Malaysia jungle occupies land up to an altitude of about 300 metres. This complex, multi storied, moist tropical rainforest dominated by dipterocarps is sometimes described as a giant green cathedral or the green jewel of Malaysia.

The floral richness of Malaysian jungle is matched by a corresponding richness of animals. It is home to many mammals and about 30 bird species found nowhere else. However, even in undisturbed conditions, large animals are relatively scarce as the jungle can be a poor source of food for grass and other vegetation eating animals, like elephants, *seladang*[59], deer and cattle that require substantial quantities of vegetation each day. To make matters even more challenging, many Malaysian jungle plants are indigestible or contain toxic chemicals.

While a high incidence of trees from the Dipterocarp family is a defining characteristic, the steamy lowland jungles host literally thousands of species crowded together, all jostling for growing space. On any hectare there may be as many as 250 tree species along with a multitude of shrubs, herbs, woody climbers and other plants. Other common trees include *jelutong*, *merbau* and *sepetir*.[60]

Moving up the slope, hill dipterocarp jungle clothes the greater part of Malaysia's inland hilly ranges from 300 to about 800 metres. Again

---

57 Referring to Charles Atlas (1892 – 1972), who became the most popular strong man of his day. He is best known for advertising campaigns featuring his name and likeness.
58 Animals especially adapted for living and moving about in trees.
59 The gaur, seladang or Indian bison is a large animal native to southern Asia and Southeast Asia. The species is considered to be threatened as the population decline in parts of the species' range is now thought to be over 70 per cent. The gaur is the largest species of wild cattle, bigger than the African buffalo and the wild water buffalo. The Malayan gaur is also called *seladang*.
60 jelutong (*Dyera costulata*), merbau (*Insia palembanica*) and sepetir (*Sindora* spp.)

species include numerous dipterocarp trees. However a major difference between the lowland and higher altitude dipterocarp forests is their plant composition. Higher altitude species of dipterocarp, such as *meranti seraya* are only found straddling ridge tops in the hill-clad jungles. *Balau* and *keruing* are also present[61]. Lowland and hill dipterocarp forest collectively account for about 18 million hectares of the Malaysian jungle.

Upper hill dipterocarp forests occupy the higher hills to around 1200 metres. The presence of a variety of dipterocarp species means that these forests continue to be called dipterocarp. Although upper hill forests have a similar vertical multi-layered structure to forests found at lower altitudes the species mix is very different. *Meranti bukit* and *keruing bukit* are both commonly found in upper hill forests.[62]

While few in number big mammals occupy a special place in the Malaysian national psyche, both culturally as well as from a wildlife conservation perspective. The *harimau* or Malaysian tiger is the national animal symbol of the country and features in the national coat of arms. This immensely powerful animal is the largest of Malaysia's cats and the most formidable of the carnivores. With an average length of 2.6 metres and weighting 150 kilograms the adult male is capable of killing buffalo and attacks on elephants are not unknown. These nocturnal animals are found in the jungles of north eastern Peninsula Malaysia.

same cannot be said for the Malaysian state of Sarawak in western Borneo. With states in the Malaysian Federation having almost complete autonomy over land use decisions jungle, management in Sarawak is entirely the responsibility of that state's administration. This is Malaysia's dark jungle secret.

For many decades the situation in Sarawak has aroused international consternation in relation to both jungle destruction and tribal human rights violations. Chapter 10 includes a discussion of the impact on and resistance of the nomadic Penan tribal people in Sarawak to uninvited logging on their ancestral lands.

---

61 meranti seraya (*Shorea curtisii*), balau (*Shorea* spp.) and keruing (*Dipterocarp* spp.)
62 meranti bukit (*Shorea platyclados*) and keruing bukit (*Dipterocarp costatus*).

*Tiger: Malaysia's national animal and largest of the country's carnivores*

Whilst jungle management in Peninsula Malaysia is reasonably sophisticated with log harvesting subject to certification requirements, the

For most of the last three decades Sarawak was governed by Chief Minister Abdul Taib Mahmud. He maintained political control over land classification and forestry licence allocation. His time as Sarawak's premier was dogged by allegations of corruption and abuse of public office. Alongside his premiership he was also Minister for Resource Planning and Environment, Minister of Finance and chaired the state's institutional investments in sectors including forestry and oil palm development. These institutions invested in and developed land in joint ventures with private sector companies such as Ta Ann Holdings and Sarawak Plantations in which members of Taib's family are major shareholders.

It is widely believed that Chief Minister Taib abused his power to benefit his family and business associates. Evidence suggests that there was systematic bribery and corruption in the process of issuing or transferring timber licences. Since Chief Minister Taib assumed office in 1981 much of Sarawak's forests have been licensed for logging or plantation development. Under Taib's tenure, Sarawak became one of the world's largest exporters of tropical timber. The state exported more tropical logs than all South American and African countries combined. In 2010 Sarawak accounted for 25 per cent of the world's exports of tropical logs, 15 per cent of global tropical sawn timber exports and almost half of all tropical plywood exports. The result for Sarawak has been an environmental, social and governance calamity.

Sarawak now has less than five percent of its intact forest remaining whilst indigenous communities who depend on the forests and land for their livelihood have become marginalised. Since no effort had been made to establish native customary rights there is doubt over the legality of forestry licences and oil palm leases issued by the government in Sarawak.

A number of international agencies have investigated and reported on jungle destruction and human rights abuses in Sarawak. Notably in 2012 Global Witness[63] reported on their research into government corruption in the logging industry and the central role played by HSBC[64] in funding and facilitating illicit decisions by the government and illegal logging activity across Sarawak.

According to Global Witness[65] HSBC bankrolled logging companies causing widespread environmental destruction and human rights abuses in Sarawak. The bank earned around $130[66] million and has violated its own sustainability policies. The bank is also providing financial services to companies widely suspected of engaging in bribery and corruption.

Global Witness also cite HSBC's track record outside Sarawak pointing to an  investigation by the USA Senate which concluded that HSBC demonstrated a "pervasively polluted culture" that allowed money launderers, drug dealers and suspected terrorists to move money into the United States financial system. In the United Kingdom a criminal investigation by the

---

63 Global Witness investigates, campaigns and reports to prevent natural resource related conflict and corruption, and associated environmental and human rights abuses.
64 HSBC (Hong Kong and Shanghai Banking Corporation) is the world's third largest publicly listed financial institution operating in 80 countries. HSBC is a British multinational banking and financial services company headquartered in London was founded in 1991.
65 For more information see: *In the future, there will be no forests left* November 2012. Global Witness, London, UK.
66 Here and elsewhere $ (dollars) means US dollars, unless the currency is otherwise specified.

Serious Fraud Office uncovered HSBC's facilitation of $22.5 million of alleged kickbacks paid by a defence firm to a member of the Saudi Arabian royal family and to public officials.

Global Witness also noted HSBC's track record of doing business with corrupt regimes and politically 'exposed' persons engaged in corruption in countries, such as Libya and Nigeria.

In Sarawak Global Witness alleged that HSBC clients played leading roles in the destruction that has left Sarawak with only five per cent of its once pristine tropical jungles. At least two of HSBC Sarawak clients are partially controlled by or owned by members of Taib's family. By providing banking services to parties notorious for endemic corruption and high level political links Global Witness contends that HSBC is at serious risk of violating international money laundering regulations.

Global Witness investigations have uncovered multiple instances of unethical, destructive and sometimes illicit operations that have facilitated illegal and unsustainable logging and palm oil plantation development. Global Witness assert that HSBC, and other major financial institutions, have a duty to ensure that their businesses do not fuel human rights abuses, environmental destruction or facilitate bribery, corruption and money laundering.

Global Witness concludes that HSBC's financial services have had a devastating effect on the world's forests and indigenous communities by helping Sarawak's logging and plantation companies pioneer and ultimately expand their destructive model of business to every major tropical forest region.

Global Witness believes that Sarawak's unscrupulous logging companies are destroying forests and people's livelihoods and could not have operated without the equally unscrupulous support from HSBC. The bank's services for this notoriously corrupt and destructive industry raise serious questions regarding HSBC's commitment to the voluntary and regulatory standards to which it purports to subscribe.

In 2011 the Malaysian Anti-Corruption Commission announced an official probe into Chief Minister Taib. It was considered that poor governance was especially pronounced in Sarawak's forestry sector that the Federal Auditor General of Malaysia described as "not satisfactory" highlighting "weak monitoring and enforcement" and "widespread infringements of licensing conditions".

Unfortunately for Malaysia Sarawak remains a serious stain on their jungle management and human rights record at a time when the country looks to throw off its developing country status. Resolution of corruption and the shame of jungle devastation and tribal human rights violations are significant issues for the country to rectify before it is will be able to advance internationally.

Adenan Satem became Sarawak's fifth chief minister in March 2014 after chief minister Taib Mahmud was forced to resign by the Malaysian Government following the long and persistent allegations of corruption and an investigation by Malaysia's Anti-Corruption Commission.

Adenan Satem was quick to accuse Sarawak's timber industry of corrupt practices. For the first time in the history of the state a chief minister confronted the timber industry.

According to Malaysian media reports[67] chief minister Adenan gathered timber tycoons of the state's logging and wood processing companies to a meeting in which he publicly accused them of using "corrupt" practices and warned them not to "mess with me."

Adenan also said that he would; "put the fear of God into people who are dishonest" and described the state of corruption in Sarawak as:

> ...very bad, a reflection of what enforcement officers have not been doing. Some, of course not all, pretend they don't know. The reason is very simple; either they are stupid, cowards or corrupt.

Responding to the chief minister's actions Lukas Straumann, the executive director of the Bruno Manser Fund[68] said:

> Today is a day of celebration for Sarawak. These are the clearest words we have ever heard from a leading Malaysian government minister to combat corruption as a root cause of deforestation and under development. We commend chief minister Adenan Satem for his courageous stance and important leadership.
>
> The international community and civil society are ready to assist Sarawak in the badly needed reforms, not only of the forestry practices, but also its governance and institutions in a wider sense.

---

67 Sarawak chief calls state's logging industry 'corrupt' Rhett A. Butler, mongabay.com November 24, 2014 – See: http://news.mongabay.com/2014/1124-sarawak-corrupt-forestry.html#sthash.CXSIVbqE.dpuf

68 The Bruno Manser Fund, a Swiss non-profit organisation campaigns on behalf of Sarawak's forests and indigenous people. Manser died in Sarawak under mysterious circumstances after a six year stay with the Penan people from 1984 to 1990. Declared legally dead in 2005, his body has never been found and foul play is widely thought to have been the cause of his death.

Chief minister Adenan said Sarawakians must not tolerate corruption anymore because millions of dollars of revenue had been lost and the state had gained a bad reputation internationally because of; "… this robbery which is carried out in broad daylight."

Adenan has initiated a major crusade to save the remainder of Sarawak's rainforests by putting a stop to illegal logging, including arming forestry department staff to combat illegal logging and corruption.

He cited illegal logging inside two national parks – Bukit Tiban and Maludam – declaring that "enough is enough," and ordered Sarawak's forestry officials to; "nail the culprits" behind illegal logging activities.

"I've seen it from the air. I've received reports on timber smuggling, illegal felling of trees and nonpayment of royalties," Adenan told reporters.

# 7

# HEART OF THE JUNGLE

## PROBLEMS AND PROSPECTS IN INDONESIA

In so many ways Indonesia characterizes the evolution and present reality of Southeast Asia. With a substantial, ethnically diverse population; a history of political turbulence; a still somewhat fragile democracy; challenges with national cohesion, and issues related to the rule-of-law, Indonesia is a work in progress. As a developing economy the country continues to place emphasis on the 'pioneering' exploitation of natural resources, including jungles where ongoing clearance to make way for intensive agriculture and mining is occurring at the expense of the biodiversity, including the survival prospects of large iconic animals.

Being home to 60 percent of the remaining jungles in Southeast Asia – equating to ten per cent of the world's tropical forests – in many ways Indonesia holds the key to the future prospects for jungle survival across the region.

As we will see over the past six to seven decades Indonesia's jungle history has not always been encouraging. However today as a result of both domestic awareness and international pressure future prospects look more promising, with serious efforts being mounted to confront long standing institutional corruption, illegal logging, forest crime, widespread land clearing and out-of-control wild fires.

Indonesia is a vast nation of 17,000 islands spanning a 5,200 kilometre arc along the equator, bridging the gap between Asia and Australia and containing about two million square kilometres of land and six million square kilometres of ocean. Indonesia is the world's largest archipelago[69] and has the world's fourth highest human population spread across 6,000 islands, ranging in size from Sumatra, Kalimantan and Western Papua to tiny atolls. The diversity of human culture rivals the plant and animal biodiversity, with as many as 400 languages spoken.

---

69 An archipelago is a term used to describe for an island group or island chain, or an area of ocean containing a large number of scattered islands.

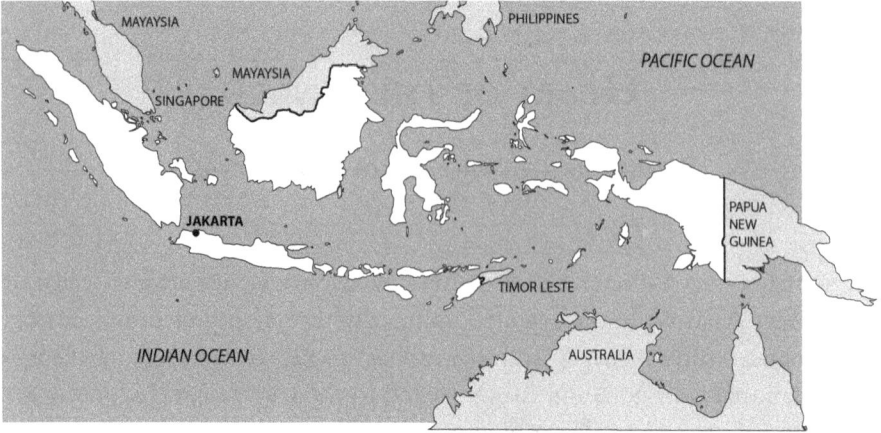

*Indonesia, the world's largest archipelago: A vast nation of 17,000 islands spanning a 5,200 kilometre arc along the equator, bridging the gap between Asia and Australia and containing about two million square kilometres of land and six million square kilometres of ocean ( JJ_Ch7_Illust_1).*

With extensive and biologically rich jungles Indonesia sits second in the world in terms of biodiversity. From the lowland jungles of Sumatra and Borneo to the snow capped peaks of West Papua, Indonesia is home to 17 per cent of all the world's bird species, 12 per cent of mammals, 16 per cent of reptiles and amphibians, and an amazing third of the world's insects.[70]

*East Kalimantan, Indonesia: rich jungles of Indonesia rank second in terms of biodiversity*

70 *Some information sourced from the Wildlife Conservation Society at www.wcs.org*

Due to its location between two continents, not surprisingly, the plant diversity of Indonesia reflects an intermingling of Asian and Australian species. The archipelago consists of a variety of habitats from northern moist lowland tropical jungles, the seasonal forests of the southern lowlands, through hill and mountain vegetation to sub-alpine zones. Having the second longest shoreline in the world, Indonesia also has many regions of swamp and coastal vegetation.

There are about 28,000 species of flowering plants in Indonesia, consisting of about 4,000 trees in 19 different forest types:[71] 2500 different kinds of orchards; 6000 traditional medicinal plants; 120 species of bamboo; over 350 species of rattan[72] and 400 species of Dipterocarps. Indonesia is also home to some unusual species, including carnivorous plants. One exceptional species known as Raffesia arnoldi is named after Sir Thomas Raffles and Dr Thomas Arnold, who 'discovered' the flower in southwestern Sumatra. This parasitic[73] plant has a very large flower, does not produce leaves and grows on a certain vine on the ground. Numerous species of insect trapping pitcher plants are also found in Borneo and Sumatra.

The dominant feature of Indonesian's flora is of course its tropical jungles – mostly seen in regions where population density is still relatively low, such as Sumatra, Borneo, Sulawesi and West Papua. On Java and Bali the vegetation is dominated by cultivated plants. Swamp forests, mangrove

---

71 For further information on forest composition and forestry in Indonesia see: *Indonesian Forestry Outlook Study* 2009 Asia-Pacific Forestry Outlook Study II, Working Paper Series, Working Paper No. APFSOS II/WP/2009/13, Centre for Forestry Planning and Statistics, Ministry for Forestry, Food and Agricultural Organization of the United Nations, Regional Office for Asia and Pacific, Bangkok 2009. This publication can be found at www.fao.org/docrep/014/am608e/am608e00.pdf
72 Rattan (from the Malay word *rotan*) is the name for the roughly 600 species of palms native to tropical regions of Africa, Asia and Australasia.
Unlike bamboo, rattan stems are solid, and most species need structural support and cannot stand on their own. Many rattans have spines which act as hooks to aid climbing up and over other plants. Rattans have been known to grow up to hundreds of metres long. Most of the world's rattan resource is found in Indonesia with the remainder in the Philippines, Sri Lanka, Malaysia and Bangladesh.
The various species of rattan range from several millimetres up to five to seven centimetres in diameter. Rattan is processed into several products to be used as materials in furniture making because of its lightweight, durability and flexibility qualities. From a strand of rattan, the 'skin' is usually peeled off to be used as rattan weaving material. The remaining 'core' of rattan is used for furniture making.
Rattans are threatened with over-exploitation, as harvesters frequently cut young stems reducing their ability to resprout.
73 A parasitic plant is one that derives some or all of its nutritional requirements from another living plant (called the host plant).

and nipa palm[74] forests are prominent along the coast.

In terms of its recent economic history Indonesia struggled to overcome the Asian financial crisis,[75] and still grapples with high unemployment, a large percentage of rural poor, ethnic and religious tensions, a fragile banking sector, government corruption, poor infrastructure, a challenging investment climate and unequal resource distribution among regions. Having said that, in recent years under more transparent governments international business confidence and investment has improved.

Keys to future economic growth of Indonesia are likely to include continuing internal reforms and building the confidence of international and domestic investors, plus strengthened environmental credentials.

Perhaps not surprisingly, given its geographic spread and domestic challenges, Indonesia has not been without its share of natural disasters and home grown turbulence. For example, in late December 2004 the Indian Ocean tsunami took 131,000 lives with another 37,000 missing, left some 570,000 people homeless, and caused an estimated $4.5 billion in damages. Terrorist incidents in 2005 and at other times have slowed tourist arrivals. Indonesia experienced several human cases of avian influenza in late 2005, sparking concerns about the possibility of a pandemic.

In addition, wildfires – seemingly impossible to extinguish – have been exacerbated by government endorsed land clearance and agricultural development. Forced transmigration programs have moved poor families from crowded central islands to less populated localities. In this program's two plus decades, more than six million migrants have been relocated to

---

74  The distinctive nipa palm, *Nypa fruticans* is a species of palm native to the coastlines and estuarine habitats of the Indian and Pacific Oceans.
75  The Asian financial crisis was a period of financial turbulence that affected much of Asia beginning in July 1997. The crisis started in Thailand with the collapse of the domestic currency. As the crisis spread, most of Southeast Asia and Japan saw slumping currencies, devalued stock markets and other asset prices, plus a steep rise in debt.

Indonesia, South Korea and Thailand were the countries most affected by the crisis, although Hong Kong, Malaysia, Laos and the Philippines were also impacted. China, Taiwan, Singapore and Vietnam were less affected, although all suffered from a loss of demand and confidence throughout the region.

The International Monetary Fund stepped in to initiate a $40 billion program to stabilise the currencies of South Korea, Thailand and Indonesia economies particularly hard hit by the crisis. In Indonesia in the wake of widespread rioting that followed sharp price increases caused by a drastic devaluation of the country's currency after 30 years in power President Suharto was forced to step down in May 1998. The effects of the financial crisis lingered through 1998.

Kalimantan, West Papua, Sulawesi, and Sumatra. Poorly educated about sustainable agricultural practices in their new homelands, many transmigrants have fared badly.

*Tsunami destruction, Acehi: Indonesia experiences its share of national disasters*

In terms of shrinking jungles, their extent across Indonesia was estimated at 145 million hectares of pristine jungle and another 14 million hectares of secondary jungle and tidal forests in 1950. This resource formed the basis of the rapid development of wood processing industries. From the late 1980s to 2000 pulp and paper industry production increased by nearly seven times, making Indonesia the world's ninth largest pulp producer and eleventh largest paper producer.

By 2001 Indonesia had lost 40 million hectares of forest during the previous 30 years or so. This is equivalent to the combined size of Germany and the Netherlands. The current rate of jungle loss is showing few signs of slowing. Between 1990 and 2010, 24 million hectares or 20 per cent of the forest area in Indonesia had been lost reducing the extent of forest cover to 94 million hectares.

With 770 species on the list Indonesia has the third highest number of threatened species in the world. However, Indonesia has the absolute highest number of threatened mammal species with 147 species – an increase of seven species since 2000.

*Orangutan, Sumatra: Indonesia has the third highest number of threatened species*

With past controversies, current and no doubt future challenges it seems fair to describe Indonesia as an 'emerging' nation. National cohesion and nation building tensions have resulted in moves to devolve government autonomy to provinces. Initiated in about 2000, this shift in policy and decision making, has led to a 'blurring' of authority in a number of areas. This trend has also resulted in lower tiers of government being financially under resourced. As a result, governance and legality have suffered from weak implementation resulting in illegal exploitation of natural resources, including forests.

Newly empowered provincial and district officials, seeking additional revenue, have authorized what may previously have been considered to be illegal activities. This has seen permits issued for poorly regulated forest concessions where over logging and subsequent land clearing have taken place.

The exploitation of Indonesia's jungle resources really stepped up in 1966 with the declaration of forests as government property under President

Suharto's regime. That proclamation was followed by the allocation of logging and mining concessions to government aligned conglomerates and politically connected families and businesses. Short concession periods, lack of regulation and 'development' incentives have encouraged over logging. Subsequent land clearing by development companies and pioneer farmers has contributed to escalating jungle loss.

In large measure, clearing has been driven by surging demand for land on which to establish palm oil plantations and other crops and by the expansion of mining for coal, gold and other minerals. The development of palm oil plantations really took off in the 1990s, with the area of plantations expanding from 1.8 million hectares in 1994 to about 2.5 million hectares by 1997.

Although the Asian financial crisis and associated political upheaval eventually lead to the downfall of the Suharto government in 1998, the exploitation of jungle resources continued and actually expanded. The subsequent government of President Bacharuddin Habibe (1998-1999) responded

*President Suharto: stepped up exploitation by declaring jungles to be government property*

to demands for reform by pushing through legislation decentralizing and devolving authority and decision making to provinces and districts. With little capacity for development planning, but a need to generate their own revenue, many local governments were quick to encourage land clearance for plantation, mining and agricultural development.

Wildfires in Sumatra and Kalimantan have become an unfortunate seasonal atmospheric and human health issue, not only in Indonesia but north into Singapore and Malaysia. These intense, devastating fires that have raged through the jungles of Indonesia over the last 20 years can largely be attributed to policy failures that have promoted exploitation and land conversion.

Fires in Kalimantan in 1982-1983 were described as both an ecological and economic disaster, damaging an estimated 46,000 square kilometres of land. While developers and pioneer farmers ignited many of the fires, the conditions that made jungles vulnerable were the result of degradation and deforestation aided and abetted by government land use policies directed at opening frontier regions to large scale commercial development.

Fires in 1997-1998 were also disastrous with more than 5.3 million hectares of forested land severely damaged in the province of East Kalimantan alone. Those fires were the result of deliberate burning by logging concessionaires, plantation owners and farmers intent on clearing land for subsequent development – often subsidized by the government.

The combined impact of deforestation and wild fires continues today. The 2009 ASEAN State of the Environment Report[76] revealed that the number of fire 'hotspots' rose to 32,400 from 19,200 in 2008. The report identified weak law enforcement, lack of supervision from local authorities and continuing land clearance as the primary cause of this increase in the Indonesian wildfire calamity.

Again in June 2013 raging fires in Sumatra blanketed Singapore and parts of Malaysia in smoke in what was considered Southeast Asia's worst air pollution incident in 16 years. Indonesian's then President Susilo Bambang Yudhoyono apologized to his neighbours promising to do everything possible to contain and extinguish the fires. Estimates suggested that the hazardous air pollution was damaging tourism and businesses in Singapore and Malaysia and could result in a bigger economic impact than the last significant fire incident in 1997 which cost the Singapore and Malaysian an estimated $11 billion.[77]

Today Indonesia's jungles continue to be degraded and cleared. Jungle extent has steadily declined since the 1960s when 82 per cent of the country was covered; to 68 percent in 1982; 53 percent in 1995, and less that 50 per cent today. The impacts of this loss of jungle cover have been widespread, including flooding and extensive soil erosion. Plus of course loss of habitat for the endangered mammals as we discuss in Chapter 9.

Logging has been a major cause of forest degradation and subsequent clearance. Indonesia is one of the world's largest exporters of tropical timber products, generating upwards of five billion dollars of export receipts a year.

---

76 *Fourth ASEAN State of the Environment Report 2009* The ASEAN Secretariat, Jakarta, Indonesia. This publication can be located at: www.asean.org/resources/publications/asean-publications/item/fourth-asean-state-of-the-environment-report-2009-2
77 As reported by Reuters in *The Dominion Post* (New Zealand) 26 June 2013.

Presently it is estimated that more than half of the remaining intact pristine jungle is already allocated to existing or planned logging concessions. Some of the logging in Indonesia is now centred on remote areas with outstanding biodiversity and wilderness values. Recent logging activity has been ramped up in previously untouched parts of Borneo and West Papua. For example, in the mid 1990s, only seven per cent of Indonesia's logging concessions were located in West Papua, but in 2012 about a fifth of all concessions were centred on that state.

According to the Environmental Investigation Agency,[78] from the late 1990s until 2005, these forests were:

> ... ransacked by organised criminal syndicates, pushing the country's illegal logging rate up to 80 per cent and resulting in the highest deforestation rate in the world.[79]

The Environmental Investigation Agency also reported that in October 2001 the Indonesian navy seized three ships carrying 25,000 cubic metres of illegal logs destined for China. This seizure proved to be the exception rather than the norm as large volumes of logs were known to have been smuggled out of Indonesia on ships destined for China. By 2004 these illegal logging operations had move east to the dense jungles of West Papua.

Environmental Investigation Agency investigations revealed the existence of an international criminal syndicate shipping illegal merbau[80] logs from West Papua to China. The syndicate comprised corrupt Indonesian government, police and military officials, plus Singapore based shipping companies and timber agents in Hong Kong and China. According to the Environmental Investigation Agency at its peak the smuggling of merbau logs reached 300,000 cubic metres a month.[81]

These revelations prompted the Indonesian government to launch an unprecedented crackdown on illegal logging in West Papua in 2005 disrupting this illicit trade. By August 2005 the price of merbau logs in China had trebled

---

78 The Environmental Investigation Agency (EIA) is an international independent campaigning organisation committed to bringing about change that protects the natural world from environmental crime and abuse.

79 *Appetite for destruction – China's trade in illegal timber.* Environmental Investigation Agency (EIA) 2012, London, UK. Pages 10-11.

80 Merbau or *Intsia bijuga* is a common hardwood tree largely from areas in Southeast Asia, Papua New Guinea, the Pacific islands, and northern Queensland, Australia. It grows to around 50 metres tall with a highly buttressed trunk. The tree has a variety of common names including merbau, ipil, and kwila.

81 *Appetite for destruction – China's trade in illegal timber.* Environmental Investigation Agency (EIA) 2012, London, UK. Pages 10-11.

to reach $700 per cubic metre as the supply of illegal logs dropped. Again in 2009, Indonesian customs intercepted 23 containers of illegal merbau logs destined for the Chinese firm Jiangsu Skyrun, a state owned company.

Currently it is considered that legally sanctioned logging concessions in Indonesia cover about 800,000 hectares of jungle a year. However, it is thought that illegal logging boosts the overall logged area to something in the region of 1.2 to 1.4 million hectares a year – possibly higher.[82] In 2004 the then Indonesian environment minister Nabiel Makarim considered that 75 per cent of logging in Indonesia is illegal. At that time and since, it has been acknowledged that despite an official ban on the export of 'raw' logs from Indonesia, logs are still smuggled to Malaysia, Singapore and other Asian countries. By some estimates the Indonesian Government is losing about a billion dollars a year in tax revenue from this illegal trade.

It seems reasonable to conclude that the Indonesia government has generally proved unable or unwilling to stem illegal logging and trading activity. This despite repeated commitments to international development assistance donors to work with them to develop policies and actions to improve forest management and governance, including stopping further conversion of forests to palm oil plantations and halting illegal logging.

*Log raft, East Kalimantan: illegal log trade costing the Indonesian Government a billion dollars a year*

82 *Appetite for destruction – China's trade in illegal timber.* Environmental Investigation Agency (EIA) 2012, London, UK. Page 7.

However all is not lost – positive changes are afoot. In May 2011 the Indonesian Government released a Presidential instruction[83] titled: The postponement of issuance of new licences and improving governance of primary natural forest and peat, as part of the country's cooperation with the Norwegian government. The Presidential instruction sought to impose a two year moratorium on new forest concession licences. This initiative is regarded as an important step towards meeting Indonesia's voluntary commitment to reduce carbon emissions.

Several issues remain to be resolved concerning the area and status of land covered by the moratorium and the amount of carbon stored in affected forests and peat lands. It has been estimated that the area covered by the moratorium is around 22.5 million hectares, comprising 7.2 million hectares of undisturbed forest; 11.2 million hectares of peat lands, and a further 4.2 million hectares of mixed category land.[84] The moratorium's application to peat lands is likely to generate the most significant environmental benefits because of the large carbon storage capacity of such areas.

The failure to include disturbed, secondary and logged forests in the scope of the moratorium is thought to represent a lost opportunity to protect, even if only temporarily, a fraction of the 46.7 million hectares of such forests still rich in carbon, biodiversity and other values.

Despite shortcomings and misgivings the moratorium is consistent with then President Susilo Bambang Yudhoyono's voluntary pledge announced in 2010 to reduce Indonesia's Greenhouse gas emissions by 26 per cent by 2020, or by 41 per cent if international assistance was provided. The Norwegian government subsequently offered a million dollars to support the President's declaration.

The announcement of the moratorium was not universally supported in Indonesia. Some in the business community, plus several parliamentarians considered that imposing limitations on forest based development opportunities would curtail economic growth and jeopardize pro employment and pro-poor development strategies. On the other hand environmental interests were disappointed by the narrow scope of the moratorium and its many exclusions and exceptions claiming that it would be ineffective in

---

83  Presidential Instruction No. 10/2011.
84  This and related information from: *Indonesia's forest moratorium. A stepping stone to better forest governance?* 2011 Daniel Murdiyarso, Sonya Dewi, Deborah Lawrence and Frances Seymour. Centre for International Forestry Research, Bogor, Indonesia.

reducing carbon emissions.

The moratorium might best be seen as a means of providing breathing space to allow the establishment of a set of conditions conducive to reducing Greenhouse gas emissions, improving forest and peat land governance and guiding the formation of a policy framework for reform beyond the two year moratorium period.

Although the moratorium is limited in time and scope, it nevertheless has the potential to facilitate improvements in forest governance. The moratorium also sends a clear message about the importance of protecting undisturbed forest and peat lands and Greenhouse gas emissions reductions.

There are also emerging signs of strengthening forest governance, including the development of an Indonesian forest log harvesting legality assurance scheme – Sistem verifikasi legalitas kayu (SVLK) – and a greater attempt to strengthen forest law enforcement, plus a better-late-than-never emphasis on corporate responsibility arising from a consumer 'push back' on environmentally unfriendly wood based products.

Indonesian company Asia Pulp and Paper has never enjoyed a great environmental reputation. By some estimates the company has cleared nearly two million hectares of tropical jungle in Sumatra since 1994, including clearing at least 200,000 hectares of high biologically diverse jungle on peat soils between 2003 and 2009.

In 2013 Asia Pulp and Paper agreed not to undertake any future logging in areas of high conservation value, or in peat land areas, and to source future log suppliers for its pulp mills predominantly from plantations. The company also agreed to get; "free, prior and informed consent" from landholders before undertaking future operations.[85]

Brokered by the Forest Trust[86] this agreement potentially represents a significant conservation advance following a decade long environmental campaign that cost Asia Pulp and Paper more than 100 customers, including Disney and toymakers Mattel and Hasbro. The agreement was also seen as a business imperative as Asia Pulp and Paper was becoming alarmed at rising

---

85  Based on article at: http://news.mongabay.com/2012/0926-tft-app-interview.html
86  The Forest Trust is a global non-profit organisation helping companies and communities deliver responsible products and assisting in the production of products that respect the environment and improve people's lives. Since its formation in 1999 a predominant focus of the Forest Trust has been to provide solutions to the issue of deforestation and the welfare of forest dependent communities.

environmental concerns about its operations by Japanese paper mills and other international customers.

"This could be a real turning point in the fight against deforestation," said Forest Trust chief executive Scott Poynton.

The agreement received qualified endorsement from Greenpeace and the World Wildlife Fund. The head of Greenpeace's forest campaign in Indonesia, Bustar Maitar agreed that green groups were; "risking our credibility" by endorsing the deal, but would; "watch and monitor closely" what happened on-the-ground.

Clearly sustainable management of jungle resources will be a key determinant of Indonesia's economic future, including defining areas to be maintained as permanent forest and the implementation of appropriate management systems for such areas. So initiatives like SVLK and the decision of Indonesian pulp and paper giant Asian Pulp and Paper, while both still young in terms of implementation, do offer the promise of improved things to come. They deserve international support and encouragement.

# 8

# PARADISE IN PERIL

## PAPUA NEW GUINEA – GOVERNANCE CHALLENGES REMAIN

Possessing tropical jungles of immense diversity, beauty and wonder Papua New Guinea (PNG) is a place and for which I have great affection and have visited numerous times.

Comprising PNG and West Papua, the island of New Guinea lies at the junction of Southeast Asia and the Pacific islands, just to the north of Australia. It is the largest contiguous area of jungle in the Asia Pacific region and after the Amazon and African Congo the third largest tropical rainforest on the planet. It features towering mountains ranges, plunging valleys, frequent earthquakes and volcanic activity. According to Professor Stephen Wurm and Shiro Hattori[87] its sharply dissected topography has helped spawn an extraordinary diversity of human cultures, including over 850 distinct languages.

The jungles of New Guinea have been ranked among the world's ten most ecologically distinctive forest regions. The complex mix of vegetation types, wide variations in climate and altitude, coupled with the long history of connection then isolation from neighbouring land masses have all contributed to a rich wildlife population with many species found nowhere else.

The extensive, diverse New Guinea jungle tract houses some 15,000 to 20,000 plant species, including 1500 trees, plus 200 mammals and 750 birds. Half of the bird species are found only on the island, including 90 per cent of the world's birds of paradise, including PNG's Bird of Paradise that graces the national flag.

---

87 *Language atlas of the Pacific Area. Part 1. New Guinea Area, Wurm S.A., Hattori., (Eds) 1983 Oceania, Australia.* Australian Academy of Humanities, Canberra, Australia.

*Lesser Bird of Paradise: graces the PNG flag*

Occupying the eastern half of the island of New Guinea, PNG is clothed in 33 million hectares of jungle – about three quarters of the country. This green mantle is both regionally and globally famous in terms of area and diversity – still seen as fierce and primitive. The astounding array of plant and animal life – six to seven per cent of the world's species – on a land area less than half of one per cent of the global land area is rated as amongst the world's most biologically diverse tropical, hilly, swamp, dry and mangrove forest communities.[88]

Perhaps historically the most jungle covered land on the planet it is of critical importance to the country's array of extraordinary tribal communities. The jungle is the foundation of their lives, cultures, social institutions and spiritual beliefs and continues to be central to their economy.

Overall PNG is sparely populated with the majority of the population continuing to rely on subsistence agriculture. However, over the past four to five decades the economic growth of the country has been underpinned by large scale mining, petroleum and logging operations.

The system of land ownership in PNG is specific to the country and a defining feature of the nation. Ninety five per cent of the land is held under a system of customary tenure, involving tribes or kinship groups, rather than individuals. This system of land ownership is recognised in the country's constitution. As the government does not own land or forests it must negotiate with land owners before it can undertake or approve activities or development. This age old arrangement is not expected to change in the years ahead, although there is an emergence of fragmentation within some tribal groupings with families seeking to be recognized as land owners in their own right.

For thousands of years the jungles of PNG have played a central role in maintaining the health and well being of tribal communities. They contribute a rich variety of foods and other items essential for daily survival, and soils for subsistence food production through the process of shifting cultivation.

More than a thousand species of jungle plants have been identified as being used by tribal communities for food, medicine, ropes, building materials, stimulants, body decoration, art, utensils and canoes. Today, valuable non timber forest products include resins, gums, meat, oils, sandalwood and rattan. Jungle related commercial activities exist for butterfly farming, insect farming, orchid production, crocodile hunting and deer, fish and cassowary farming.

Increasingly throughout the years, but more so since the end of the Second World War, jungles have been utilised for their timber and have

88 For more details see: *Papua New Guinea Country Study on Biological Diversity*, Sekhran, N. and Miller, S.,(eds) 1994 Department of Environment and Conservation, Port Moresby, PNG and *Loggers, Donors and Resource Owners: Papua New Guinea Country Study*, Filer C., and Sekhran N., 1998 IIED Policy that Works for Forests and People Series No. 2. Meyers J., (ed.)

become a major source of revenue for tribes and the government. Some 15 million hectares of forests are accessible for logging. The bald facts are that about 25,000 hectares of jungle are cleared and another 125,000 hectares logged annually. By 2002, about three million hectares of forest had been degraded by logging. Such logged areas are vulnerable to further degradation and clearing.[89] The fate of logged forests varies – while some have been left substantial areas have subsequently been cleared, burned or converted to palm oil plantations.[90]

Also since the Second World War tribal groupings have wanted development to take place to improve their access to basic goods and services. Pressure has been exerted on the government to organize timber concessions and attract timber companies. Tribes have sometimes taken it upon themselves to seek out potential developers and subsequently approach the government to endorse development proposals.

As I have indicated, I have visited PNG on a number of occasions and been able to see jungle first hand, talk to government officials and read a good number of reports. Extensive deforestation and forest degradation has occurred over the decades from about the 1970s. The activities driving these changes across much of the accessible parts of the country have been commercial timber production, land use conversion, the development of mines, palm oil plantations and agriculture.

Various commentators and reports assert that there are few mechanisms in place in the form of rational land use planning, enforcement of low impact logging practices or control over erosion and water pollution to ameliorate jungle destruction and to protect vulnerable plant or animal habitat. Equally, there is little evidence of effective reforestation or ecological restoration of degraded forest areas being carried out.

The compelling conclusion from an analysis of jungle change over the past thirty or more years is that forestry regulation has not produced useful results for longer term sustainable forest management or biodiversity conservation.

Much of the country's logging operations are carried out by Malaysian

---

89 Data used is based in part on: *The State of the Forests of Papua New Guinea*, Shearman, Phil L., Bryan, JE, Ash J., Hunnam P., Mackey B, and Lokes B., 2008. University of Papua New Guinea, Port Moresby, PNG.
90 *Forest conversion and degradation in Papua New Guinea 1972-2002.*2009 Shearman P.L., Ash J., Mackey B., Bryan J.E., Lokes. *Biotropica* 41, 379-390.

timber companies with logs being shipped to Chinese wood product manufacturers and then exported around the world as manufactured products.[91] The timber industry is dominated by a single multinational company, Rimbunan Hijau, based in Sarawak, Malaysia. Rimbunan Hijau controls approximately 60 per cent of PNG's timber industry and owns a number of PNG businesses, including The National, one of only two national newspapers. Rimbunan Hijau has frequently been singled out for criticism.

The State of the Forests of Papua New Guinea[92] has some strong statements about the treatment of PNG forests, including:

> The changes being wrought by the degradation and clearance of the nation's forests will drastically reduce future options for the country's economy and its capacity to maintain a healthy and content society. In an era when the majority of nations are striving to regain lifestyles, resource use practices and economic activities that adhere to the principles of sustainable development, PNG has watched over the wholesale destruction of its natural capital, with little or no benefit accruing, even short-term, for the great majority of its people.

It is apparent that the longer existing forest management practices continue, the more difficult it will be to introduce changes directed at sustainability and averting the widespread collapse of jungle based economic activity.

The PNG Government has developed several pieces of legislation incorporating administrative procedures to acquire legal approval from customary land owners for logging. Prior to 1991 the government was able to acquire rights for the development of forest resources through provisions set out in the Forestry (Private Dealings) Act 1971.[93] This legislation enabled the government to obtain the rights over forest resources area and to issue

---

91 See China's Impact on Papua New Guinea's Forest Industry. Bun Y., King T., Shearman P.L. 2004 Forest Trends (www.forest-trends.org).
92 For a summary of the report see: http://www.scienceinpublic.com.au/blog/wp-content/uploads/State-of-Forests-of-PNG_Concise.pdf
93 The Forestry (Private Dealings) Act 1971 granted customary land owners the right to apply to have their forests declared a Local Forest Area (LFA) and to sell their timber subject to the approval of the PNG Forestry Minister. This Act bypassed the timber rights purchasing (TRP) procedures that had previously governed all logging activity and timber sales. For more information refer to: What can be learnt from the past? A history of the forestry sector in Papua New Guinea 2007 Papua New Guinea Forest Studies 1, Overseas Development Institute, London, UK. Website: www.odi.org.uk

permits to timber companies.[94]

The operation of this timber rights purchase system gave rise to many allegations of corruption and malpractice. A commission of inquiry into the forest industry headed by Justice Barnett was established in 1987 by the then Prime Minister, Paias Wingti to investigate these allegations. The damning findings of the Commission's report prompted a complete legislative overhaul in the forestry sector.[95]

The Barnett Report documented numerous instances of corruption and serious abuse of power by politicians and officials. The inquiry was considered to be the most devastating anti corruption inquiry into the forestry industry ever undertaken in PNG and uncovered many instances of money changing hands between Asian businessmen and people in positions of authority in PNG.

The findings of the inquiry provided compelling evidence that some timber companies were operating illegally. A much quoted phrase from the inquiry's report describes what can only be described as an appalling situation:

> It would be fair to say, of some of the companies, that they are now roaming the countryside with the self-assurance of robber barons; bribing politicians and leaders, creating social disharmony and ignoring laws in order to gain access to, rip out and export the last remnants of the province's valuable timber...

In another section of his first report, Justice Barnett said the former Forests Minister, Ted Diro, was using a timber company, Angus (PNG) Pty Ltd, to raise funds offshore for the use of his political party. The judge said his investigations had shown:

> ... that the Angus timber operation raised questions of ministerial impropriety, corruption, transfer pricing and the abuse of government policy of a very serious nature.

The report went on to say:

> A major concern... is the evidence of blatant corruption at high levels of government and the practice of ministers and senior public servants of negligently, and sometimes deliberately, ignoring and contravening the laws

---

94 *What can be learnt from the past? A history of the forestry sector in Papua New Guinea. Papua New Guinea Forest Studies 1*, 2007 Overseas Development Institute, London, UK. Website: www.odi.org.uk

95 1989 *Report of Commission of Inquiry into Aspects of the Forestry Industry (The Barnett Report)* TE Barnett, Commission of Inquiry, Port Moresby (seven Interim Reports and Final Report).

of Papua New Guinea's Parliament and the policies of its government.

Judge Barnett said his investigations into the allocation of timber resources and operations on the island of New Ireland indicated that bribery, corruption and the buying of support were so widespread they had become a major social sickness. He said speedy decisions on forestry matters by ministers in both national and provincial governments and by public servants without bothering to consult other authorities or to check the facts, and without regard to due legal process raised serious suspicions. These suspicions, he said, were confirmed by oral and documentary evidence that in some cases was overwhelmingly strong.

Justice Barnett reported that in the absence of any coherent policy a de facto policy had evolved that:

> ...was resulting in some of the country's best forests being mercilessly ripped apart and the owners cynically ripped off. Such formulated policy that does exist is frequently and openly contravened by ministers, senior officers, inspectors and timber operators. The result is that the national department has lost its sense of purpose and provincial forestry officers are wandering leaderless with no sense of belonging to a profession of foresters.[96]

Judge Barnett's reports were a blunt, brutal and frank assessment of the state of the timber industry at the time. His conclusions and findings certainly made international headlines and gained the attention of the authorities from the Prime Minister down. His reports and findings also had major implications for the way international agencies approached the forest industry in PNG from the 1990's onwards.

Justice Barnett called for a slowdown in logging activity; the reform of national forest policy; the establishment of a nationally integrated forest service; the development of improved consultation procedures in the allocation of timber harvesting permits and the formalisation of detailed requirements for sustained yield forestry practice.

He indicated that because of the state of the industry at that time, including transfer pricing, the PNG government had missed out on a substantial amount of revenue. A report prepared for the United Nations Centre on Transnational Corporations estimated that the PNG Government was receiving approximately half of the average revenue received by

---

96 For a detailed account of the Barnett enquiry and findings, plus an overview of PNG's history since Independence see Sean Dorney's outstanding book: *Papua New Guinea - People Politics and History since 1975*, 2001 ABC Books Sydney, NSW, Australia

the Philippines for its logs, and one third of what was received by the Government of Sabah in Malaysia.[97]

As a result of Justice Barnett's inquiry, the PNG Taxation Office recovered several million kina[98] from a number of major timber operators to settle backdated tax claims.

In the aftermath of the Barnett inquiry the PNG government invited the World Bank to implement reforms in the country's timber sector as part of a large structural bank loan. However, dissatisfied with the slow pace of reforms, the World Bank began withholding loan payments in 1996. A new World Bank loan was subsequently negotiated, along with a temporary ban on new logging concessions, plus actions intended to improve forest sustainability; reduce impacts on biodiversity; increase community participation, and strengthen environmental impact assessment and monitoring processes. This revised loan was suspended by the PNG government in 2003.[99]

Also in response to the Barnett Report, a national forest policy was published in 1990 and a new Forestry Act passed in 1991. The existing timber rights purchase mechanism was replaced with forest management agreements that extended over a 30 year period. The objectives of the new policy and Act were to manage and protect the nation's forest resources, both as a renewable natural asset and source of economic growth and employment. These government initiatives were also intended to achieve greater national participation in the forest industry with increased domestic processing of harvested logs. The PNG Forest Authority was subsequently established with a mandate to implement the new policy and supporting legislation.

While the regulatory framework for forest management in PNG has been considerably strengthened since the Barnett Report the success of many reform measures remains problematic. Politically connected corporations and individuals often flout or circumvent reforms designed to improve forest

---

97 For a lot more detail see: *The Barnett Report* Commission of Inquiry Interim Report Number 4, Volume 1 page 85 "New Ireland", plus *Transnational Companies and Global Forest Resources -* An assessment for WWF UK produced as a submission to the Commission on Sustainable Development 1994 by Nigel Dudley.
98 A PNG kina is worth around 33 cents US.
99 *A serious case of conditionality: the World Bank gets stuck in Papua New Guinea* 2004 Filer C., *Devel Bull* 65, 95-99.

governance. Such chronic abuses are a key reason why PNG is perceived to be among the most corrupt nations in the world – ranked 144th out of 175 nations by Transparency International in 2013.[100]

This situation has lead to several countries continuing to question forest management activity in PNG. For instance in a letter on 27 January 2005 the United Kingdom Timber Trade Federation warned its members not to purchase timber originating from PNG (and the Solomon Islands) suggesting there was little evidence to provide even a minimum guarantee of legality. The trade federation concluded that timber originating from PNG and the Solomons should be deemed 'very high risk'.[101]

In March 2001, a PNG Forest Authority economist concluded that:

> Notable problems within the industry include: still virtually no sustainable forestry projects; poor logging practices with little compliance to the Logging Code of Practice; widespread environmental damage; very few long-term benefits, causing social upheaval; corruption a persistent problem at all levels of the industry; minimal domestic processing investment, and; many proposed projects too small to be viable.[102]

These observations are supported by more recent independent reviews of the forestry sector concluding that many of the problems of the forest industry identified in the Barnett Report still exist. The major problems in the PNG logging industry can be summarised as non compliance with laws in relation to forest allocation, forest operations and timber permit conditions. Across PNG, logging operations have frequently resulted in negative social and ecological impacts, and been contrary to both the country's constitution and the long term economic, ecological, cultural and security interests of PNG and its citizens.

The benefits to landowners from logging operations, whether as royalties or in the form of infrastructure development are frequently considered not to be commensurate with the real economic value of the resource and are often short lived. A 2006 World Bank review of the socioeconomic and financial impacts of timber permits concluded that:

> Few lasting benefits are reaching landowners because payments to the poorest and most remotely located communities are too small and

---

100 Transparency International (www.transparency.org) and *the Legality Compliance Toolkit* 2014 Australian Timber Importers Federation.
101 For further and related information see: Partners *in crime: The UK timber trade, Chinese sweatshops and Malaysian robber barons in Papua New Guinea's rainforests* Greenpeace 2005, London, UK.
102 PNG Forest Authority internal report, 2001.

ephemeral to have a lasting impact and are not complemented by investment in public services by government. Payments that reach rural populations, furthermore, are primarily used to purchase consumables by men and infrequently invested.[103]

Despite serious deficiencies that have been identified in forest management practices, the PNG government has continued to defend large scale logging of forests under concession arrangements. Indeed, the government has sought to accelerate the granting of timber permits, further stretching the capacities of government forest authorities. In a review of forest harvesting projects the Independent Forestry Review Team concluded that:

> …by attempting to respond to the political call for more new forestry projects quickly, the National Forest Service has initiated far more new project developments than it has the capacity to process properly.[104]

An International Tropical Timber Organisation study of the forestry sector found that:

> The more significant issues are to do with the compliance of the government itself with the laws of PNG when deciding to designate a forested area for logging purposes; negotiating the agreement with landowners; managing, monitoring and enforcing the agreement; and when extending current agreements. It is believed that the narrow focus of the PNGFA [PNG Forest Authority] on exploitation of the forest resource for the primary financial benefit of the national government presents a conflict of interest which colors decisions made by the government at all levels.[105]

Rather than increasing local employment and skills by producing sawn timber, veneer and plywood, log exports provide only modest economic benefits for the country and traditional landowners. For example, local communities are typically paid just $11 a cubic metre for kwila[106] logs, a valuable timber that typically fetches $240 per cubic metre on delivery as raw logs to wood product manufacturers in China.[107]

103 Forestry and conservation project, 2006 World Bank, Rural Development and Natural Resources Sector Unit, East Asia and Pacific Region.
104 *Review of Forest Harvesting Projects Being Processed Towards a Timber Permit or a Timber Authority: Observations and Recommendations Report.* Independent Forestry Review Team, 2001. Report to the Inter-Agency Forestry Review Committee. Port Moresby, PNG.
105 *Achieving the ITTO objective 2000 and sustainable forest management in Papua New Guinea.* Report of the diagnostic mission. International Tropical Timber Council, April 2007
106 Kwila (*Intsia bijuga*) grows in Southeast Asia, the Philippines, Solomon Islands, Fiji and Papua New Guinea. It is called merbau in Indonesia.
107 *China's Impact on Papua New Guinea's Forest Industry.* Bun Y., King T., Shearman P.L. *2004* Forest Trends (www.forest-trends.org).

Compounding matters, in May 2010, the PNG Parliament stripped the country's many communal groups of key land right protections in an effort to increase certainty for resource developers by minimising project delays arising from court injunctions. As a result local communities are now unable to impede projects that are considered environmentally risky and are likely to find it more difficult to sue offending corporations for environment damage.

Forests continue to play a significant role in the economic development of PNG being among the top three sectors in terms of financial contribution to foreign exchange earnings and to the economy. Forest industry companies provide direct employment to over 10,000 people, mostly in the rural communities, they also built infrastructure, such as roads and bridges, and provide a source of income to tribes through timber royalties and developmental levies.

In common with many other countries PNG has put in place a system of protected areas, although the proportion of land set aside in this way is one of the lowest of any country. This protected areas program has been supported by government agencies, land owners and local communities, scientists, non-government conservation and development organisations, plus international donor agencies.

This conservation initiative has been motivated by recognition of the fundamental role the country's extensive jungles play in maintaining the extraordinary biological diversity of PNG. The contribution such areas to the livelihoods, food production, culture and customs of Papua New Guineans is also recognized as is the vital ecological services they provide.

According to the impressive policy paper by Bill Laurance and others,[108] the formula for addressing shortcomings by the PNG Government should have three elements. First log exports should be sharply curtailed while overall log production reduced to more sustainable levels. The paper estimates the sustainable level to be 25 to 30 per cent of the present harvest of approximately 2.5 million cubic metres a year. The paper also observes that a number of tropical nations have curtailed or fully banned log exports

108 *Predatory corporations, falling governance, and the fate of forests in Papua New Guinea* 2011 Laurance William F., Kakal Titus, Keenan Rodney J., Passingan Simon, Clements Gopalasamy R., Villegas Felipe, and SodhiNavjot S., *Conservation Letters* 4 95-100.

in order to reduce timber theft and corruption and to promote their own domestic wood processing industries.

Secondly the policy paper prescribes that PNG should foster the development of domestic value adding wood processing industries – that are tiny by international standards – by establishing a more favourable investment climate for 'downstream' processing.

Finally PNG should reinstate the legal rights of traditional communities to challenge foreign corporations for past or anticipated environment damage. The amendment that stripped communities of these fundamental rights places environmental safeguards solely with the Environmental and Conservation Minister – an action widely considered to be unacceptable.

I have not sought to deliberately single out PNG for harsh comment in this chapter. However, the examination of this country – central to the tropical jungle story – serves as an illustration of on-the-ground governance challenges faced by governments and communities across much of Southeast Asia. In the specific case of PNG the future looks more promising. It is apparent that the current government is showing some willingness to better manage the country's precious jungle resources for the economic, social and environmental benefit of tribal communities and the country as a whole.

Bringing forward the intended log export ban from 2030 to 2020 that was announced by the Minister for Forests, Douglas Tomuriesa in 2015 was applauded as a bold step in the right direction. Already this initiative has encouraged present log importers and others to give serious thought to domestic wood processing with attendant employment, regional development and overall economic benefits for the country.

# 9

## SYMBOLS AND PRIDE ENDANGERED

### TIGER, ELEPHANT, ORANGUTAN AND RHINO UNDER THREAT

As you will have gathered, whilst this book is essentially about jungles from a plant – or more precisely tree perspective – it is the animals, especially the high profile large mammals that give jungles much of their intrigue, majesty and excitement. It is the iconic wild animals of Southeast Asia that attract attention and rally people to the cause of 'saving-the-jungle'. These animals also generate national pride and symbolism right across the region.[109]

Certainly the emblematic large animals are central to the passion, biodiversity and romance of Southeast Asian jungles. And it is the clearing, fragmentation and degradation of their jungle homes that in large measure has placed these animals on the critically endangered list.

I do not intend to track through all the animals – I hope a sample will illustrate the points I want to make. So let us look at the tiger, orangutan, elephant and rhinoceros – they all epitomise Southeast Asia – and while regarded with patriotic pride are all under siege.

The message in this chapter should be clear from the outset – protect their jungle homes and the animals will stand the greatest prospect of survival. At the very least such action should be motivated by a recognition that these animals play a vital role in national identity, pride, commercial 'branding' and tourism – so their care makes complete economic sense. Manage the jungles of Southeast Asia better and both the animals and economy will thank you for it.

Unfortunately, in human-animal 'interaction' encounters animals invariably come out the losers. Elephants are relocated in Peninsula Malaysia when they continue to traverse traditional migratory routes now planted with palm oil; orangutans perish when jungle is removed for land use

---

109 The content of this chapter has been greatly assisted by a number of books, including, *100 Animals to see before they die*, 2007 Nick Garbutt with Mike Unwin Bradt, Connecticut, USA, plus several websites including the World Wildlife Fund www.wwf.org.au

development, and tigers are badly impacted by factors, such as fragmentation, illegal hunting and decline in prey species.

To give just one example of the impact of a single event on population decline before we head off into this chapter, take the catastrophic deforestation of the Brahmaputra River watershed caused by floods that overwhelmed Kaziranga National Park[110] in India[111] in 1998. In a disastrous 48 hours almost all of the park's population of three quarters of the world's remaining Indian rhinoceros,[112] as well as elephants, tigers, buffalos, deer and wild pigs were swamped by flood waters up to ten metres deep.[113]

Park officials reported that 32 rhinos had drowned. It was also reported that about half the surviving rhinos were forced on to higher ground outside the park boundary where they were almost immediately attacked by poachers. Hundreds of deer and wild pigs were also drowned, while again poachers killed many of those escaping to drier ground.

So now to turn to the iconic large mammals of the Southeast Asian jungles – let us first consider the tiger – one of the most revered animals on the planet.

To get some sense of the tiger in Southeast Asia we will start by looking across the past and present range of this magnificent animal. As we will see the current status and future survival prospects for wild tigers is bleak.

---

110 Located in the Indian state of Assam, the park is a World Heritage Site hosting two thirds of the world's remaining Great One-horned rhinoceroses. Kaziranga also boasts the highest density of tigers among protected areas in the world. The park is also home to large breeding populations of elephants, water buffalo and swamp deer. The park is crossed by four major rivers, including the Brahmaputra.

111 Yes, not quite in the Southeast Asian scope of this book, but an example nonetheless.

112 The Indian rhinoceros (*Rhinoceros unicornis*) is also called the greater one horned rhinoceros and Asian one-horned rhinoceros. Among terrestrial land mammals native to Asia, the Indian rhinoceros is second in size only to the Asian elephant. This heavily built species is also the second largest living rhinoceros. Listed as a vulnerable species this animal is found primarily in north eastern India's Assam state and in protected areas in Nepal. The Indian rhinoceros range has been drastically reduced by excessive hunting. Numbers in the wild today are thought to be less than 3000.

113 *Major Indian Rhino Catastrophe*, Myanmar Forestry Journal, Vol. 4, no. 3, July 2000, Page 23.

Today, across their range, tigers[114] face unrelenting pressure from habitat loss, poaching and retaliatory killings. They are also forced to compete for space with growing human populations. In the face of this pressure – if concerted action does not occur – wild tigers may never again roam the Earth, but be confined to zoos or to live only in stories, pictures and myths.

*Tiger: current status and future survival prospects in Southeast Asia are bleak*

---

114  The number of tigers in the 1900's – over 100,000 – dropped to 4,000 in the 1970's. Today the five remaining subspecies (Bengal tiger, Indochinese tiger, Siberian tiger, South China tiger, and Sumatran tiger) are all on the critically endangered species list.

The Bengal tiger or Royal Bengal tiger, roams a wide range of habitats, including high altitudes, tropical and subtropical rainforests, mangroves, and grasslands. They are primarily found in parts of India, Nepal, Bhutan, Bangladesh and Myanmar.

Indochinese tigers are located across southern China, Vietnam, Malaysia, Cambodia, Laos, Thailand and eastern Myanmar. It is estimated that fewer than 1,200 Indochinese tigers are left in the wild.

The Siberian tiger or Amur tiger is considered a critically endangered species with the primary threats to its survival in the wild being poaching and habitat loss from intensive logging and development.

The South China tiger is the smallest of all the tiger subspecies, and it is the most critically endangered. Little is known about their exact numbers in the wild, but some estimates would put the number at under 20 tigers. The reality is that no South China tiger has been seen in the wild for the last 20 years.

The Sumatran tiger is found on the Indonesian island of Sumatra. Their habitat ranges from lowland forest to mountain forest and includes evergreen, swamp and tropical rain forests. It is estimated that only between 400-500 Sumatran tigers remain in the wild.

But let us back up for a minute and to start with a poem – I have always liked the first and last verse of William Blake's[115] poem *The Tyger.*

Tyger! Tyger! burning bright
In the forest of the night
What immortal hand or eye
Could frame thy fearful symmetry?

A 'big cat' of beauty, strength and majesty, the tiger is one of the world's truly magnificent animals and one of the most endangered. The harsh reality is that tiger numbers in the wild are at an all time low. About 95 per cent of wild tigers have been lost in just over a century.

The largest of the big cats, tigers rely primarily on sight and sound, rather than smell. Unlike lions, leopards and cheetahs, tigers prefer to live in areas where they can hide among trees, scrub and tall grasses, camouflaged by their dark stripes, and ambush their prey.

On average, tigers give birth to two-to-three cubs about every two and a half years. Cubs generally gain independence at two years of age and attain sexual maturity at three to four years for females and at four to five years for males. Juvenile mortality rates are very high with about half of all cubs not surviving beyond their first two years. Tigers in the wild have been known to live to an age of 26 years.

Males of the largest subspecies, the Siberian tiger may weigh up to 300 kilograms. For males of the smallest subspecies – the Sumatran tiger – the upper weight range is around 140 kilograms. Within each subspecies, males are heavier than females.

Although now critically endangered, tigers once occupied a vast region of wilderness that extended as far north as Siberia, as far south as the Indonesian island of Bali; as far west as Turkey, and east to the Russian and Chinese coasts. From icy cold mountains and forests to steamy, tropical jungles, tigers have adapted to a wide variety of habitats.

Young tigers live with their mothers until they are old enough to fend for themselves and find territories of their own. Determined mostly by the availability of prey species territories can range from a few hundred to many thousands of hectares. Formidable predators, tigers have razor sharp claws, long teeth and powerful jaws. They typically hunt alone and can bring down animals far heavier than themselves, including buffalo, deer and wild boar.

---

115 An English poet, painter, and printmaker William Blake (1757-1827) is considered a seminal figure in the history of the poetry and visual arts of his time.

*Tiger and cub: now critically endangered tigers once occupied a vast region of wilderness*

Of the eight original subspecies of tigers, three have become extinct in the last 60 years. On average one every 20 years – the Bali tiger extinct in the 1930s; the Caspian tiger in the 1970s, and the Javan tiger in the 1980s. Despite these trends wild tigers still exist in Eastern Russia, China, Vietnam, Cambodia, Lao Peoples' Democratic Republic, North Korea, Thailand, Malaysia, Indonesia, Myanmar, Bhutan, India and Nepal.

Up to two thirds of remaining tigers – between 2,700 and 4,300 – are Bengal tigers, found in India, Bangladesh, Nepal, Bhutan and Myanmar. About an additional 350 live in captivity.

The Sumatran tiger is now found only on the Indonesian island of Sumatra where their habitat ranges from lowland jungle to mountain forest. It is estimated that only between 500 to 600 Sumatran tigers remain in the wild, but the actual number may be as low as 400.

The Siberian tiger is also severely threatened. In 1991, one third of the Siberian tigers were killed to meet the demand for their bones and other body parts used in traditional Chinese medicine. Only 150 to 200 survive in the wild, on three reserves in the Russian Far East. An additional say 500 are managed in international zoo conservation programs.

Little is known about the status of the Indochinese tiger due to its scattered habitat across Thailand, as well as Myanmar, southern China, Cambodia, Lao, Vietnam and Malaysia. It is estimated that 900 to 1,200 are left in about 75 isolated reserves. About 60 are found in zoos.

Returning to the Sumatran tiger of Indonesia, of these about 400 are found in five national parks and two reserves. Perhaps 100 live outside reserves and their habitat may well be lost to expanding agricultural development in the near future.

Despite 20 years of international effort, we are still losing ground in the battle to save the tiger. As the mountains, jungles, forests, and tall grasses that have long been their home disappear so too do tigers. Poachers continue to poison water holes or set wire snares to kill tigers, selling skins and body parts as ingredients in traditional Chinese medicine.

If this isn't bad enough, agricultural expansion, new roads, human settlement, industrial expansion and hydroelectric dam developments continue to push tigers into smaller and smaller areas of jungle. These habitat fragments are often surrounded by rapidly growing, human populations.

Southeast Asia's explosive population growth requires increasing areas of land to be converted to agriculture. For example, Indonesia has the same population as the United States, but only ten percent of the land area. Almost all of Indonesia's lowland jungle has been cleared for rice cultivation and other agricultural crops. In addition to food, local communities also need to use surrounding patches of jungle for livestock grazing and firewood.

As tigers compete with humans and industry for land, they find less-and-less to eat. Local people frequently hunt the same prey as do tigers, pressing tigers to sometimes resort to preying on domestic animals and on rare occasions humans. This can result in threatened villagers poisoning, shooting, or snaring encroaching tigers. However, perhaps the single greatest threat of extinction that looms over most Asian wildlife, especially the tiger, is the massive demand for traditional Chinese medicine.

The consumption of traditional remedies made from wild animal parts, such as tiger bone, bear gall bladder, rhinoceros horn, dried geckoes and other animal parts is phenomenal. The use of tiger parts is not new, but in recent years the increase of the standard of living across Southeast Asia has made these remedies available to a greater percentage of the population. Today it is believed that at least 60 per cent of China's billion plus inhabitants use some traditional medicines.[116] Clearly strengthening economies and improved personal incomes of Southeast Asian citizens has caused an accelerated demand for traditional medicines and for prices to soar, lifting the trade in wildlife products to an estimated $6 billion a year business.[117]

The rising demand for tiger parts and increasing prices continues to be an irresistible incentive for poachers. Since China has almost eradicated its own tiger population it is now looking for new supplies of tigers from Bangladesh and Nepal. The World Wildlife Fund estimates that one third of breeding age female tigers from Bangladesh and Nepal was lost between 1989 and 1991.[118]

Even though, along with Nepal, Japan, South Korea and Thailand, China is a member of the Convention on International Trade in Endangered Species (CITES) the laws are widely ignored.

Russia has also become a key supplier in the tiger trade due to political, economic and social instability. Poaching a single tiger can bring in ten years of income on the black market. It is estimated that in 1991 one third of the Siberian tigers were killed to supply traditional Chinese medicine markets.[119] In the more austere economic times of the post Soviet era, poaching has become a serious threat to tiger survival across the old Soviet Union.

Because the demand for tiger products continues to grow, and with poaching still prominent in India, Russia and Southeast Asia, additional measures are needed to curb both the demand and supply of tiger body parts. When combined with habitat protection and the promotion of alternatives to traditional Chinese remedies improved national legislation and international support are vital parts of the strategy to save the tiger.

To survive in the wild tigers need large tracts of habitat with sufficient

---

116 Traditional Chinese medicine at: http://en.wikipedia.org/wiki/Traditional_Chinese_medicine
117 TRAFFIC (the wildlife trade monitoring network). See www.traffic.org/trade/
118 http/worldwildlife.org
119 For additional information see website: www.tigersincrisis.com/trade_tigers.htm

water to drink, animals to eat and cover for hunting. Optimal tiger habitat includes a core area of at least 1,000 square kilometres free from most human activity. Without wilderness wild tigers will not survive. On-the-ground security is essential to protect tigers from poachers. Enforcement officers, park guards and staff need to be funded, trained, organised, equipped and legally empowered to protect the tiger from illegal hunting.

As long as a demand and market for traditional Chinese medicine thrives, lucrative returns will provide incentive to breaking the law. For this reason, in addition to laws banning the sale and trade of tiger parts, medicinal alternatives need to be developed and promoted. When combined with efforts to protect tiger habitat, these actions can help reduce the economic and political circumstances that continue to undermine attempts to save the tiger.

The political and economic problems that jeopardise the ability of the endangered tiger to survive are large and complex. It is difficult to provide immediate solutions to poverty, corruption and the global pressure on jungle resources and land. Clearly legally forcing relatively poor communities to choose between their own livelihood and the survival of the tiger is not a realistic solution. Despite efforts the prospect of losing the last of the wild tigers within the next decade looms large.

To be effective, laws to protect the tiger need to be reinforced by education that puts the case for conservation. Economic resources are needed to support underfunded enforcement efforts, as well as community based programs on tiger compatible development. Obviously the bottom line for the survival of tiger populations in the wild is adequate, quality habitat. So getting right down to it tigers like other iconic Southeast Asian animals showcased in this chapter need a home in which to live that meet their long term survival needs.

The orangutan is perhaps the ultimate wildlife emblem of Southeast Asian jungles. Its compelling facial expressions and thoughtful, emotion filled eyes have instant appeal. The orangutan ranks amongst our closest relatives. Genetically they are about 97 per cent identical with us – they are intelligent, thoughtful and inventive. Now only in Borneo and Sumatra – the red apes, as they are sometimes called – possess a culture and a sense of beauty. Their name is composed of the Malay words for person (*orang*) and forest (*hutan*) that means *person of the forest*.

The orangutan is the world's largest tree dwelling animal and the only great ape found outside Africa. Their forearms are 30 per cent longer than their legs and both the hands and feet are equally adept at gripping. They cannot jump like many other primates, but instead span gaps by swinging a tree back and forth until they can reach adjacent branches.

*Borneo orangutan: the world's largest tree dwelling animal and the only great ape outside Africa*

The first orangutans were carried off to Europe in 1776 as shaggy curiosities for the scientific collection of the Dutch prince, Willem V. One of the first orangutans to reach United Kingdom was a female named Jenny who was exhibited in a London zoo in 1837.

A young Charles Darwin paid her a visit in the spring of 1838. Darwin had returned from his circumnavigation in the *Beagle* in October 1836 and was working on his seminal work: *On the origin of species by means of natural selection or preservation of favoured races in the struggle for life* in 1859 and in 1868 his treatise on *The Descent of Man*. After he had seen Jenny, Darwin wrote in his diary:

> ... let man visit the ourang-outan in domestication – see his intelligence –
> man in his arrogance thinks himself a great works – more humble, and I
> believe true, to consider him created from animals.

A descendent of Jenny was presented to Queen Victoria and the young monarch observed the ape with horror stricken fascination typical of Victorians. Later she described her impression as; *frightful, and painfully and disagreeably human.*

However, it is only in the Western countries that the orangutan is called by this name. At home they are called "mawas" or "maia", "kahiyu" and several other local names.

There is some strange history surrounding the scientific name of the orangutan. Initially it was labelled *Simia sativa* as the first orangutan arriving in Europe came from Angkola in the north of Sumatra. Scientists thought the animal must have come from the African country

*Charles Darwin: controversially at the time argued that humans were created from animals*

of Angola and consequently took it for a chimpanzee. As a result of this confusion the scientific name was wrong. *Pongo pygmaeus* is now used. Modern science distinguishes two species – the Borneo orangutan *Pongo pymaeus* and the Sumatran orangutan *Pomngo abelii*. Scientists also list the Borneo orangutan as having three geographically distinct subspecies.

While the other great apes spend a substantial part of their time on-the-ground, the red ape rarely climbs down from the jungle canopy. Orangutans

*Young Queen Victoria: horrified that orangutans were "disagreeably human"*

possess a heavy set body and a thick neck, with a large head. At birth their face is rosy, but later becomes dark brown to leathery black. Their facial expressions are very much like those of humans. There is thoughtfulness and intelligence in their large dark eyes – referred to as innocence and wisdom combined. Orangutans have long rough hair of different shades, depending on the species and also often differing amongst members of the same genetic group – bright orange, rusty red, reddish brown and even brown black like gorillas.

Male orangutans reach a head to rump length of about 100 centimetres and a standing height of about 135 centimetres. They can weigh up to 70 kilograms. The females are markedly smaller weighing about 40 to 50 kilograms.

*Indonesian orangutan: thoughtfulness and intelligence in their eyes*

With arm spans of two and a half metres and more as well as long hands and feet orangutans have a great capacity for moving hand-over-hand at dizzy heights through the tree tops. Although at first glance rather awkward looking, they have tremendous physical strength.

Females become sexually mature and often have their first child when they are about 15 years old. Pregnancy lasts about eight and a half months. Young orangutans depend on their mothers for a longer time than other animals and receive tuition for about seven years. This includes everything young growing orangutans need to know – from climbing to botany. The longest childhood in the animal kingdom goes hand-in-hand with a very low fertility rates. On average only one baby about every eight years, so females orangutans have no more than three children.

Because their menu comprises several hundred kinds of fruit and because during low fruit yielding periods they feed on leaves, the orangutan appears to possess a detailed botanical understanding. Orangutans may be familiar with at least 1000 plants – edible, inedible, poisonous and beneficial. Some suggest that the botanical repertoire of orangutans could extend to 4000 plants.

Orangutans have few enemies – the Sumatra tiger, dogs and wild pigs and of course, humans. In Indonesia they have been hunted and eaten probably for thousands of years. Human demand for young orangutans escalated dramatically in the 1970s when it became fashionable in social circles, and among the officer ranks in the Indonesian armed forces, and police to show off with one of the *oh-so-cute* animals as a member of the household.

But young orangutans grow up and go from cute to calamitous as far as the household furniture, fittings and other inhabitants are concerned. So eventually they get either locked away in a backyard cage or sold to an orangutan butcher for meat.

Orangutans as domestic pets and toys is bad enough with bleak long term prospects for these temporary status symbols. However, in some instances it seems that cruelty is carried to greater extremes where these good natured apes are forced to be circus clowns or kept as sex slaves – shut up in brothels and used in porn movies.

However, it was the destruction of the orangutan's habitat, particularly the explosive expansion of palm oil plantations that has led to what has been called genocide on a massive scale. Borneo really started to lose orangutan jungle habitat in a big way from the 1980s. According to reports, between

1985 and 2005 about 850,000 hectares of jungle was cleared each year. In total this amounted to about seventeen million hectares. By late 2008 about half of Borneo had been deforested.

So now the remaining living space of the approximately 50,000 orangutans still to be found in the Indonesian provinces of West, East and Central Kalimantan, as well as in Sabah and Sarawak, is extremely fragmented. Many red apes now live like castaways on an island with populations confined to these remaining 'sanctuaries'.

The situation in Sumatra is even more serious than on Borneo with 70 per cent of the jungle cover already gone, including about 90 per cent of lowland rainforests – the preferred habitat of the orangutan. During the fifteen years between 1990 and 2005 the number of red jungle kings of the island decreased by half and continues to shrink. By the beginning of 2006 it was estimated that only about 7,000 were left.

The elephant is our largest land animal. Asian elephants, *Elephas maximus* show clear differences from their African cousins. They are smaller, reaching a height of perhaps a little more than four metres and a weight of around five tonnes. Female Asian elephants are smaller again.

The Asian elephant is still found in 13 countries on both the Asian mainland and on a number of islands. Asian elephants are divided into three sub species.[120] Since 1986 the Asian elephant has been listed as endangered as the population has declined by at least 50 per cent over the last three generations – about 75 years. The species is seriously threatened by habitat loss, degradation and fragmentation. In 2005 the wild population was estimated at between 41,000 and 52,000 individuals.

Asian elephants were formerly widely distributed south of the Himalayas, throughout Asia into China as far north as the Yangtze River and across Southeast Asia. Today they exist in parts of India, Sri Lanka, Myanmar, Thailand, Laos, Vietnam, Malaysia, Sumatra and northern Borneo.

---

120 The three sub species of the Asian elephant are:
*Elephas maximus maximus*, or the Sri Lankan elephant found on the island of Sri Lanka is the largest of the Asian elephants. Sri Lankan bull elephants appear less predisposed to growing tusks.
*Elephas maximus sumatranus*, or the Sumatran elephant is found on the Indonesian island of Sumatra. It is the second smallest of the Asian elephants.
*Elephas maximus borneensis*, or the Borneo elephant, was only classified as a separate sub species in 2003 after tests revealed differences in the genetic make-up from mainland elephants. The Borneo elephant is the smallest of the sub species and sometimes referred to as the pygmy elephant. These elephants are found in Borneo.

*Asian elephant: once widespread but now confined to parts of India, Sri Lanka, Myanmar, Thailand, Laos, Vietnam, Malaysia and Borneo*

The Asian elephant has smallish square ears and relatively smooth skin that is often densely freckled. Its back is not as sloping as the African elephant, the head rather than the shoulders is the highest part of the body, the trunk has a single finger-like projection rather than two, and the hind foot has four nails not three.

The elephant's trunk is actually a long nose serving many functions. It is used for smelling, breathing, trumpeting, drinking and also for grabbing things – like food. Another distinguishing characteristic of the Asian elephant is that the forehead has two hemispherical bulges, as distinct from the flat front of the African elephant. Unlike African elephants that rarely use their forefeet for anything, Asian elephants are more agile using their feet in conjunction with their trunk and head for moving objects.

The elephant's incisor teeth develop into tusks that grow throughout the animal's life. Deeply rooted in the cranium, tusks can grow two and a half metres and can weigh over 60 kilograms each. In Asian elephants, only the male has long tusks, although some males have no tusks at all. Females have short tusks – called *tushes* – usually hidden under their upper lip

The only other teeth elephants possess are four molars – two on the upper jaw and two on the lower jaw – which are replaced seven times throughout their lives. As the animal ages and the teeth wear away, they are replaced by the next set. If an elephant lives long enough to have used up all of its seven sets of teeth tragically it will starve to death.

*Asian elephants: distinct differences from their African cousins*

Elephants use their tusks to dig for roots and water, strip bark from trees and to fight each other. An adult elephant can consume up to 130 kilograms of food a day. They roam over great distances while foraging for the large quantities of food they require to sustain their massive bodies.

Because of their size and need for enormous quantities of food and water, elephants require large habitats. They also drink at least once a day and can consume 150 litres at a time so they never stray very far from a water supply.

Female elephants, or cows, live in family herds with their young, while adult males, or bulls, tend to be solitary. Having a calf is a serious commitment as elephants have a longer pregnancy than any other mammal – almost 22 months. Cows usually give birth to one calf every two to four years. At birth, elephants already weigh some 90 kilograms and stand about a metre tall. Females in the herd act as foster mothers assisting with play and babysitting duties.

Asian elephants inhabit grasslands, jungles and other localities, such as secondary forests, scrublands and agricultural areas. Over this range of habitats they are seen from sea level to over 3,000 metres. Elephants are often found in open grassland at the jungle's edge as they prefer areas that combine grass, low woody plants together with taller trees.

*Asian elephants: often found in open grassland as they prefer areas that combine grass and low woody plants together with taller trees*

Adult cows and calves move about as groups although adult males disperse from their mothers upon reaching adolescence. While generally living alone, bull elephants may form 'bachelor groups' from time-to-time. Cow calf groups tend to be small, typically consisting of three or four related adult females and their offspring. However, it is not uncommon to see larger groups containing as many as 15 adult females. These groups are led by the oldest female – the group matriarchal – she directs the herd's daily routine and movements in search of food and water. Herds may disperse into subgroups that maintain contact through low frequency, long distance vocalizations.

Bull elephants will fight each other to get access to females, although fierce contests are rare. Bulls reach sexual maturity around the age of 12 to 15 years. Between the age of 10 and 20 years, bulls undergo a yearly phenomenon known as *musth* – when they can become extremely aggressive. This is a time when the testosterone[121] levels in bull are much greater than during *non musth* periods.

Asian elephants are generally timid and much more likely to flee than attack. However, solitary rogues are frequently an exception to this rule and can make unprovoked attacks. It is always dangerous to approach cows with calves. When an Asian elephant makes a charge, it tightly curls up its trunk and attacks by trampling its victim with its feet or knees, or if a male, by pinning it to the ground with its tusks.

You know the line about: *an elephant never forgets*, well elephants are considered to be highly intelligent. Asian elephants have been captured from the wild and domesticated for thousands of years. These most powerful of animals have been employed to move heavy objects, such as logs, carry humans in religious ceremonies and even in wars. They have been used for their ability to travel over difficult terrain and have served as mobile hunting platforms.

Elephants have long been an integral part of Asian culture and religion. In Indian society elephants were considered to bring good luck and prosperity. Hindus revere elephants as the god *Ganesha* and even today elephants are central to many religious festivals. Also elephants play an important role in Buddhist beliefs where they are recognised for their steadfastness and as symbols of physical and mental strength.

---

121 Testosterone is a male sex hormone and is secreted in the testicles of males and also the ovaries of females. In male's testosterone plays a key role in the development of male reproductive tissues as well as promoting secondary sexual characteristics, such as increased muscle and bone mass and the growth of body hair.

Habitat loss has hastened the decline of the Asian elephant. Vast areas of jungle have been converted to agriculture and isolated elephant populations have become cut off from their traditional seasonal migratory routes. This has resulted in increased elephant-human conflict. Poaching for ivory continues to threaten long term elephant survival.

Now finally let us turn to the rhinoceros. Of the three Asian rhinoceros species – Sumatran, Javan and Indian – only one is showing signs of conservation success and increasing numbers.

At the beginning of this chapter I outlined the dramatic loss of Indian rhinos when the Brahmaputra River flooded overwhelming the Kaziranga National Park resulting in drowning and poaching of a substantial proportion of the park's rhino population. However, the Indian, or greater one horned rhino, is enjoying some conservation success. Its original range extended from Pakistan all the way through India, Nepal, Bangladesh, Bhutan and Myanmar. However, in 1975 only 600 remained. Decades of conservation efforts have seen the population rise to 2,500 by 2007, and a reclassification from endangered to vulnerable. Today there may be about 3,000 surviving in the wild. Unfortunately this success story is outside the Southeast Asian focus of the book.

*Indian rhinoceros: signs of conservation success and increasing numbers*

Across Southeast Asia habitat loss and the ongoing threat of poaching continue to push rhino populations to the brink of extinction. Southeast Asian resident rhinos – named after the islands on which they are found – Sumatra[122] and Java rhinoceros are seriously endangered due to rampant poaching and loss of lowland jungle habitat. Poached for their horns, the rhinos are caught with snares, pit traps, poisoned or electrocuted. Asian rhino populations are now distressingly small.

Although hunting is now illegal and the use of rhino horn in traditional Chinese medicine is banned in most countries, Sumatran and Javanese rhinos are among the world's most endangered large animals. The use of rhino horn still persists, despite the clear medical evidence that rhino horn has no medicinal or therapeutic benefits whatsoever. Since the 1970s the demand for horn, especially from China to treat a range a variety of ailments ranging from epilepsy, fevers and strokes to AIDS has increased.

Southeast Asian rhinoceroses live in thick jungle where they are hard to see. They are smaller than the other types of rhinoceros making it easier for them to move in dense jungle habitat. Both species can swim and generally live in places where there is a good water supply. They wallow in mud to keep cool and for protection from the sun and biting insects. Both species tend to live alone except at breeding time.

Rhinos are browsers feeding on shoots, twigs, leaves and fruit. They use their jaws and chest to bend trees so they can feed from branches. They have large heads, small eyes and solidly muscled bodies. They also have excellent hearing and a keen sense of smell – useful attributes in dense jungle habitat that more than compensates for their poor eyesight.

Although Sumatran rhino numbers exceed those of their Javanese relative, it is considered to be more threatened due to a combination of habitat loss, continuing poaching and population fragmentation. The population of Sumatran rhinos has declined by 70 per cent over the last two decades. No single Sumatran rhino population is estimated to have more than 75 remaining individuals, making them extremely vulnerable to extinction due to natural catastrophes, diseases, poaching and further habitat loss.

The smallest of living rhinoceroses the Sumatran rhino – sometimes also called the lesser two horned rhino or hairy rhino – once ranged across much

---

122 There are two surviving subspecies of Sumatran rhino; the Western Sumatran rhinoceros (*Dicerorhinus sumatrensis sumatrensis*), and the Borneo (or eastern) Sumatran rhinoceros (*Dicerorhinus sumatrensis harrissoni*).

of Southeast Asia and beyond. They occupied the foothills of the Himalayas existing in Bhutan and eastern India, through Myanmar, Thailand, possibly to Vietnam and China, and south through the Malaysian Peninsula to the islands of Sumatra and Borneo. Numbers are thought to have at least halved between 1985 and 1995. Today, the population is estimated at fewer than 200 individuals, located in small pockets on Sumatra, Peninsular Malaysia, and Borneo. The Borneo population is considered a distinct subspecies, numbering perhaps fewer than 25 animals. Until recently the Sumatran rhinoceros was also found in Malaysia Peninsula. However, there have been no recent sightings there and this population may well have become extinct.

The Sumatran rhino is characterized by fringed ears, reddish-brown skin coated with long hair and wrinkles around its eyes. Calves are born with a dense covering of hair that turns reddish and becomes sparse, bristly, and in older animals almost black. Females are thought to be territorial and to avoid each other. They give birth to a single calf every three to four years. Calves are born from October to May corresponding to the rainy season. Calves gain independence at about one and a half years and may join other juveniles before taking up a solitary existence. Females are thought to reach sexual maturity at six to seven years while males reach sexual maturity at ten years. Life span is thought to be similar to other rhinos at around 35 to 40 years.

Unlike the situation for orangutan survival outlined in Chapter 14, there may not be definitive studies, but observations suggest that the Sumatran rhinos also inhabit logged jungle areas where there is sometimes an increased abundance of ground cover vegetation and regenerating plants. However, on the debit side of the ledger the construction of logging roads and tracks makes the jungle more accessible to poachers.

The critically endangered Javanese rhino is probably the rarest large mammal in the world. Also known as the lesser one horned rhino, the species historically roamed from north eastern India through Myanmar, Thailand, Cambodia, Lao, Vietnam, and the islands of Sumatra and Java in Indonesia. Today, no more than 50 individuals are thought to survive in the wild. There are none in captivity.

Only one subspecies of the Javanese rhino remains and is restricted to

Ujung Kulon National Park,[123] Java, Indonesia. The Vietnamese subspecies of Javan rhino was pronounced extinct after the last remaining individual was found dead with its horn removed.

So, no doubt about it – this has been a despondent chapter – with the large animals that give Southeast Asia so much of its character and identity pretty much teetering on the brink of extinction. Clearly the region and its jungles will be so much the poorer both biologically and economically if any of these high profile symbols of the region slide over the brink into oblivion.

While I have showcased the tiger, orangutan, elephant, and rhinoceros their precarious future survival prospects typify other larger mammals in Southeast Asia, such as the clouded leopard, the proboscis monkey and a couple of gibbon species. All are at risk as their jungle homes contract and people pressures increase.

*Endangered Proboscis Monkey: future survival prospects at risk as habitat contracts*

123  The Ujung Kulon National Park is located in the extreme south western tip of Java and includes the Ujung Kulon peninsula and several offshore islands. In addition to its natural beauty and geological interest, particularly for the study of inland volcanoes, the park contains the largest remaining area of lowland jungle in Java containing several species of endangered plants and animals, including the Javan rhinoceros.

So what can be done to prevent this occurring? It is so much about sustaining their jungle homes, but also about stamping out the scourge of poaching for the traditional Chinese medical trade – a practice that has nothing to do with medical science, but everything to do with ancient superstition and witchcraft. We must end this superstitious nonsense for the sake of Southeast Asia's large, precious and irreplaceable jungle dwelling wild animals.

# 10

## IT'S STEALING — PLAN AND SIMPLE

### ILLEGAL LOGGING THREATENING JUNGLE SURVIVAL

Yes – it's very easy to be upbeat when talking about the wonder of Southeast Asia's tropical jungles – their diversity, magnificence and beauty. However, there is a dark and traumatic downside. There is nothing magnificent or beautiful about the extent, speed and finality of jungle destruction. Driven by clearance, or *deforestation*, and illegal logging this destruction is having a catastrophic impact on jungles, their habitat quality and their biodiversity.

It is not my purpose in putting pen-to-paper to be overly pessimistic, although I need to highlight the very serious survival threats confronted by Southeast Asian jungles. Having done so, we can then turn to ways that might assist in dealing with these threats.

It is all too clear that the 'drivers' of biodiversity loss and subsequent species extinction are jungle clearance, degradation and fragmentation, along with the impact of introduced species.[124] Breaking the jungle up into discrete areas restricts plants, and particularly animals to habitat 'islands' often too small to support healthy breeding populations. Also bad news is the impact of illegal logging and forest clearance on the welfare of indigenous communities and rural poor. All together it is far from a pretty story.

We will see in this chapter that the combined impacts posed by jungle clearance, both for large scale palm oil and tree plantations and for small scale farming, plus unsustainable or illegal logging contribute to a whittling away of the Southeast Asian jungle estate. Roading and track access related to logging operations frequently provide a 'gateway' through which migrant rural poor and illegal loggers gain easy access. Also it is much easier to marshal an argument to remove remaining trees and convert the land to some other use if the jungle has already been damaged and devalued.

---

124 For further information see: *Parks, People, and Policies: Conflicting Agendas for Forests in Southeast Asia* Kathy MacKinnon. In *Tropical Rainforests: past, present, and future.* Eldredge Bermingham, Christopher W Dick, and Craig Moritz (eds.) 2005 The University of Chicago Press, London.

*Jungle clearing: a principle reason for biodiversity loss*

It is now compellingly clear that frequently working in tandem illegal logging and deforestation threaten the survival of some of the richest and more diverse habitats on the planet – nowhere else is this more apparent than in the majestic tropical jungles of Southeast Asia.

Stepping back for a moment to look at the situation globally, in 1990 it was estimated that every year some 170,000 square kilometres of tropical jungle were being cleared. This is an area equivalent to the size of Cambodia. Deforestation figures for Indonesia indicate that 18 million hectares of forest were lost between 1985 and 1997.[125] Of the world's three major tropical forest regions, Southeast Asia has the highest annual rate of jungle destruction.[126]

Since 2004 the Environmental Investigation Agency has been conducting investigations into illicit timber flows in a number of producer countries, including Indonesia, Myanmar, Russia, Laos, Mozambique, Madagascar and

---

125 *Indonesia: Environment and natural resource management in a time of transition.* 2001 World Bank, Washington, DC.
126 *Reflections on the tropical deforestation crisis.* Laurance, WF 1999 Biological Conservation 91: 109-117

China. The findings from these investigations highlight the impact of illegal logging to supply the Chinese market; the destruction of jungle tracts; loss of revenue for developing countries, and rampant corruption and conflict.

Illegal logging underpins a multi-billion dollar global trade in stolen timber. A study by Interpol and the United Nations Environment Programme[127] reported that up to 30 per cent of annual timber production, and a massive 50 to 90 per cent of timber harvested in tropical countries may be illegal, creating perhaps a $100 billion global trade. A 2010 study[128] estimated that more than 100 million cubic metres of illegal timber may be traded annually.

Illegal logging thrives in countries with poor forest governance and where transparency and law enforcement are weak. Such problems are common in developing countries with natural resource focused economies, so these nations experience the worst impacts of illegal logging. Illegal logging causes developing countries to lose an estimated $15 billion a year in revenue and taxes.[129]

In the densely populated countries of Southeast Asia, jungle loss is continuing at an unprecedented rate, both to meet the needs of the land hungry growing population and to fuel economic development. The estimated figures for loss across Indonesia alone have almost trebled over the last 12 years from 900,000 hectares annually in 1990 to at least 1.7 million hectares by 1997[130] and closer to 2.5 million hectares by 2006. Clearly such rates of loss have an enormous impact on biodiversity and on restricting future development options.

Yes, a substantial contributor to deforestation in Southeast Asia and elsewhere is poverty – communities attempting to eke out a living felling trees for fuel wood and cash, and clearing trees to plant food crops. There is really only one long term solution to this situation – poverty eradication – and herein lie the future prospects for much of the world's remaining jungles.

Illegal logging has and continues to have a debilitating impact on hunter

---

127 Interpol/United Nations Environment Programme, Green Carbon, Black Trade: Illegal Logging, Tax Fraud and Laundering in the Worlds Tropical Forests, 2012
128 Chatham House, *Illegal Logging & Related Trade: Lessons from the Global Response*, 2010
129 World Bank, Justice for Forests, 2012
130 *Where have all the forests gone?* Holmes, D 2002 EASES Discussion Paper, World Bank, Washington, DC.

gatherer[131] life styles of tribal jungle dwelling people who have occupied parts of Southeast Asia for tens of thousands of years. For example, for decades indigenous tribal people in Sarawak have been protesting about uninvited logging on their ancestral land. In a bid to stop the destruction of their lands thousands of people formed human barricades across logging roads.[132] While these human blockades have not halted logging activities they brought the plight of these ancient hunter gatherer tribal communities to world attention. Hundreds of media reports about the blockades and sympathetic international community's support put the spotlight on the Malaysian government to do something to reign in indiscriminate, illegal, destructive logging deep in the jungles of Sarawak.

Sarawak tribal communities have suffered the devastating effects of logging since their jungle homes have been licensed out as timber concessions. The construction of fences and log blockades across logging roads was an attempt by traditional jungle owners to stop further destruction of their ancestral lands. These blockades were not carried out without sacrifice on the part of tribal people, especially the nomadic Penan. Many had to stop gathering food to guard blockage sites. Others had to walk great distances to show their solidarity, leaving women and children to fend for themselves.

In the case of the Penan – possibly the worse affected by the ferocity of jungle exploitation – their existence in Sarawak stretches back to the dawn of humanity. The land and jungle are the most important economic resource they have and the relationship between the jungle and the community has been the cornerstone of their culture and society.

The jungle also has deep significance in the spiritual life of tribal communities. This reverence for the land meant that it could not be brought or sold. This principle is enshrined in customary law. Customary land tenure allows anyone in the tribe to cultivate and use the land.

The 1988 Right Livelihood Awards awarded by the School of Peace Studies at the United Kingdom University of Bradford included among the three recipients the Sahabat Alam movement in Sarawak. Their award

131 A hunter gatherer community is one in which most food is obtained from wild plants and animals, in contrast to agricultural societies which rely mainly on domesticated plant and animal species.
132 For a lot more detail about the flight of tribal people in Sarawak see: *The Battle for Sarawak's Forests*, 1989 World Rainforest Movement – Sahabat Alam Malaysia.

citation[133] noted:

> ... their exemplary struggle to save the tropical forests of Sarawak, the
> home of its tribal peoples, from destruction.

The citation also recognised the courage of S. Mohamed Idris, founder
and President of Sahabat Alam Malaysia plus Harrison Ngau and the Penan
people:

> ... who at great personal risk have inspired and led the fight against the greed-
> driven demolition of one of South-east Asia's greatest remaining rainforests.

The problems of Sarawak's tribal people were considered substantial and
getting worse – soil erosion, pollution of critical water supplies, lack of
traditional food and emerging health problems.

The issue of illegal logging is now receiving international scrutiny. Let us
be crystal clear, no one should condone illegal logging. In Southeast Asia
illegal logging robs governments of income that should come their way via
log royalties and taxes. If the land is tribally owned, as it is the case in PNG,
traditional owners are denied income from their ancestral assets and forests
are irreparably degraded, especially where protected areas like national parks
are illegally logging.

Plain and simple, illegal logging is stealing logs. That is, felling trees and
extracting logs without requisite approvals under the relevant laws of the
country in which the forest in question is situated. In particular the laws
that relate to the approvals needed for the harvesting of logs, or conversely,
violating laws that explicitly prohibit the felling of trees and the removal of
logs from particular areas, such as national parks, nature reserves or tribally
owned land.

Illegal logging and the trade in stolen timber are among the most
destructive of environmental crimes and directly threaten jungle ecosystems.
As already outlined, illegal logging is often a precursor to clearance and
conversion to another form of use.

The financial flows generated by illegal logging operations have
exacerbated armed conflicts in Southeast Asian countries. Violence and
murder are often associated with the illegal timber trade, with loggers,
journalists and local activists being targeted.

---

133 Press release: *1988 Right Livelihood Awards shared by environmental activists, community architect and torture-
rehabilitation doctor,* The Right Livelihood Award, School of Peace Studies, University of Bradford.

*Soldiers from the US 32nd Infantry Regiment intercepting illegal timber being smuggled through the Narang Valley in Afghanistan's Konar province: no one should condone illegal logging*

The scale and threat posed by illegal logging has prompted concerted actions in many parts of the world. Such measures appear to be having an impact as a 2010 Chatham House[134] analysis found that global illegal timber production had reduced by 22 per cent since 2002. Improved law enforcement in key producer countries is largely credited with this reduction, along with the development of legislation and market reforms in consumer countries that have sought to prohibit trade in illicit timber and to promote legal trade.

Further, in consumer counties like the USA, across Europe and in Australia, the sale of timber products linked to illegal logging can put downward pressure on prices of legitimate timber products. Also

---

134 Chatham House is an independent policy institute based in London that brings together people and organisations with an interest in international affairs. Chatham House is considered to be a world-leading source of independent analysis, informed debate and influential ideas on how to build a prosperous and secure world.

importantly, the hint of illegal logging runs the risk of denting the compelling environmental attributes of timber products. Climate change abatement, low energy manufacture and renewability credentials are central aspects of a commendable marketing push to expand the use of timber products. Having the prospect of timber from illegal logs ripped out of a national park, or the habitat of an endangered species is damming for an industry that has such a positive environmental message to underpin its marketing efforts.

A leading catalyst for this anti illegal logging effort north of the equator was the decision of the United Kingdom Government not to allow timber products to be purchased for government sponsored projects where legality was raised as an issue. As this was about a third of all timber product sales the timber industry – lead by the United Kingdom Timber Trade Federation – sat up and took serious notice. This government initiative was followed by the larger building suppliers and hardware retailers also demanding proof of legality for the timber products they sourced and sold.

I am not going to get into a lot of technical nitty gritty here, but the European Union has developed their Forest Law Enforcement, Governance and Trade (FLEGT) action plan that has been passed by the European Union Parliament and rolled out across Europe in 2013. The USA took a short cut in 2008, and simply extended the 1900 *Lacey Act* to cover wood based products.

The *Lacey Act* was first placed on the USA statute books to prohibit the trade in wildlife, fish, and plants illegally taken, possessed, transported or sold. The original Act was directed more at the preservation of game and birds by making it a crime to poach game in one state with the purpose of selling it in another. The legislation was also concerned with the potential problems of introducing exotic species of birds and animals into American ecosystems. Since about the 1960's The *Lacey Act* has been used more aggressively to apprehend and prosecute smugglers of reptiles into the USA.[135] The 2008 Amendment to the *Lacey Act*, which prohibiting trade in illegal timber, established the concept of 'due care' as a core mitigation incentive and required importers to file declarations on the nature and origin of timber products imported.

The European Union's 2003 FLEGT Action Plan includes two components: Voluntary Partnership Agreements – establishing licensing

---

135 For example see: *The Lizard King*, Bryan Christy, 2008. Twelve Hachette Book Group USA, New York.

systems for verified legal timber imported into the European Union from partner countries – and the European Union Timber Regulation, that from March 2013 banned the import of illegal timber. The actions in the United Kingdom, Europe, the USA and in Australia via the *Illegal Logging Prohibition Act 2012* all involve the introduction of measures to restrict the importation of illegally logged timber products.

In anticipation or in response to the amendments to the *Lacey Act*, Europe's FLEGT procedures and Australia's *Illegal Logging Prohibition Act* many major timber importing companies, trade associations and building supply retailers have initiated work on developing more stringent timber product procurement measures. Such actions are a positive response to illegal logging at the consumer country end of the timber product supply chain. However, at the sharp end of supplier country practices illegal logging continues to be a malignant cancer.

As an example, and taking advantage of the work of others, an investigation into rosewood smuggling in the Koh Kong province of Cambodia was carried out by Denis Gray.[136] His report in *The Jakarta Post* on 27 November 2012 provided a sobering case study of illegal logging activity in Southeast Asia – and of the powerful criminal forces, weak governance, poor law enforcement and lucrative nature of this illicit trade.

Denis Gray reported that a force dubbed the 'Rambo Army' tried, but could not stop illegal logging gangs armed with military weapons. Also that activists were sometimes killed in Southeast Asian countries, citing the example of one of Cambodia's leading environmentalists, Chut Wutty who was killed while investigating illegal logging of highly prized rosewood.[137] Illegally logged rosewood smuggled into China is turned into furniture that sells for hundreds of thousands of dollars. This illegal logging has increased dramatically in recent years and driven the region's rosewood to the brink of extinction.

---

136 Denis Gray is an accomplished investigatory journalist and is the Bureau Chief for Associated Press in Bangkok.
137 Rosewood is the popular name given to several species of tropical trees that have brownish with darker veining timber. All are strong and heavy, taking an excellent polish and are much in demand for furniture, luxury flooring, turned items, guitars and other high value applications. Rosewood is found in Cambodia, southernmost China, East Timor, Indonesia, Malaysia, PNG, the Philippines, the Solomon Islands, Thailand and Vietnam. It can also be found in northern Australia and in western Pacific Ocean islands.
  *Pterocarpus indicus* is the more common species of rosewood that also has a number of local names, including Pashu Padauk, Malay Paduak, New Guinea Rosewood and Narra that can refer to several *Pterocarpus* species.

The jungle region of Koh Kong in southwest Cambodia, where most villagers earn less than two dollars a day, is a hub of rosewood illegal logging. Locally, illegally felling a rosewood tree is considered to be better than winning the lottery. In 2011 a cubic metre of top grade rosewood could be sold for up to $3000 to middlemen who hover around forests and construction sites of dams and roads in Thailand, Laos, Myanmar and Vietnam.

Even though law enforcement capacity is limited, thousands of illegally felled trees have been seized in recent years and many of those accused of involvement in the trade have been arrested, including a son of a Cambodian general and twelve Thai police officers.

"The spectrum of illegal rosewood logging ranges from loggers, military and police officers to forestry officials. This network runs the industry," says Chavalit Lohkunsombat, who commanded the Rambo Army, and remains head of the forest protection unit of Cambodia's Nakhon Sawan province.

Faith Doherty of the Environmental Investigation Agency[138] described the illegal logging activity in Southeast Asia as a "war" igniting violence between officials and smugglers and sometimes among rival gangs. "This is not just an environmental issue. It drives corruption and criminal networks. There is a lot of violence and blood spilled before the rosewood ends up in someone's living room," she said.

The Environmental Investigation Agency estimates that nearly 50 Cambodian loggers and smugglers have been killed and others arrested over the past two years in clashes related to illegal logging activity.

Once the smuggled rosewood snakes its way to furniture makers in China, often via Vietnam, the price escalates. A sofa and chair set of high quality *hongmu* or rosewood can sell for $320,000 according to an article in the Monterey County's *The Herald*.[139] A four poster bed was seen by the Environmental Investigation Agency with a price tag of a million dollars.

Chinese customs documents show Cambodia exported 36,000 cubic metres of logs to China from January 2007 to August 2012. Although the Cambodian government issued a blanket denial it is a different story on-the-ground. According to foreign conservationists and the Cambodian human rights group LICADHO,[140] with investigators in Koh Kong, there

138 *Appetite for destruction – China's trade in illegal timber.* Environmental Investigation Agency (EIA) 2012, London, UK.
139 *Rare hardwood sparks gunfights, corruption in Asia* 24 November 2012, Denis D. Gray. www.montereyherald.com
140 Cambodian League for the Promotion and Defense of Human Rights.

is a substantial amount of 'tree laundering' occurring. They say logging companies falsified documents to make it appear that their wood came from permitted areas when it was actually harvested from other locations.

LICADHO assert that Cambodian military police trucks ferry the illicit timber to warehouses in remote areas of Koh Kong. It is then shipped down the Tatay River headed for Vietnam, or by road to the capital, Phnom Penh then across the Vietnamese border.

The volume of rosewood consumed by China alone suggests that most was obtained illegally. Documents show that China's appetite for rosewood is soaring – from just 66,000 cubic metres in 2005 to 500,000 cubic metres in 2013. The dramatic growth in the wealth of the Chinese middle class is cited as the main reason for the surge in demand for rosewood and the 'driver' of illegal logging activity in adjacent Southeast Asian countries. A number of Chinese websites offer rosewood furniture. Some makes its way to USA, European and other Western markets.

In 2009 and 2011, USA authorities raided the Tennessee plants of the Gibson Guitar Corporation seizing $500,000 worth of imported ebony and rosewood that was to be used to make guitar fingerboards. In August 2012 Gibson paid $350,000 in penalties to settle federal charges of illegally importing ebony, but rosewood was not part of the charges.

The Environmental Investigation Agency says that to curb the trade in illegally logged rosewood products Southeast Asian nations must push for rosewood to be included in CITES,[141] the international treaty protecting trade in endangered flora and fauna. Rosewood species from Madagascar and Brazil are already listed. But as a final note of caution LICADHO suggests that even such action may prove too late for forests. "The rosewood is almost all gone from Koh Kong after just a few years. It has been a total rape," says LICADHO's In Kongchit.[142]

Have no doubt, illegal logging goes well beyond Southeast Asia. In an article titled: Mahogany's Last Stand, Scott Wallace reports on the reality that illegal logging has all but wiped out Peru's mahogany trees.[143] Investigations revealed that illicit practices account for three quarters of the annual Peruvian timber

141 CITES is the Convention on International Trade in Endangered Species of Wild Fauna and Flora. It is an international agreement between governments aimed to ensure that international trade in specimens of wild animals and plants does not threaten their survival.
142 The World Post See: http://www.huffingtonpost.com/huff-wires/20121124/as-asia-rosewood-wars/
143 National Geographic Vol 223, No 4 April 2013.

harvest with most exports going to the USA, but with increasing volumes destined for Asia. Scott says:

> Despite a crackdown on mahogany logging that began five years ago and a sharp decline in production, much of the timber reaching markets in the industrialized world is reported to be of illegal origin.

In another development Interpol's first major international operation against illegal logging and forest crime in February 2013[144] resulted in numerous arrests and the seizure of millions of dollars worth of timber, and some 150 vehicles across South America. Officials in Bolivia, Brazil, Chile, Colombia, Costa Rica, Dominican Republic, Ecuador, Guatemala, Honduras, Paraguay, Peru and Venezuela carried out inspections and investigations on vehicles, retail premises and individuals, as well as surveillance and monitoring at ports and transport centres.

Law enforcement officials reported a total of 194 arrests, with 118 individuals under investigation. Seizures of timber and other forest products were estimated at more than 50,000 cubic metres, enough to fill 2,000 trucks, with a total value of about eight million dollars.

Manager of the Environmental Crime Programme at Interpol David Higgins said the operation:

> Marks the beginning of Interpol's effort to assist its member countries to combat illegal logging and forestry crime, which affects not only the health, security and quality of life of local forest dependent communities, but also causes significant costs to governments in terms of lost economic revenue.

He added that one of the key aims of the Interpol operation was the development of practical cooperation and communication among national environmental law enforcement agencies, forest authorities, police, customs, and specialised Interpol units.

Interpol criminal intelligence officer Davyth Stewart said the intelligence gathered during the operation will be used as a foundation for more:

> … incisive actions against illegal logging to be taken by Interpol, in cooperation with its member countries. Interpol will continue to support countries to establish long term sustainable improvements in law enforcement *responses to illegal and unsustainable deforestation.*

---

144 Environmental News Service, 20 February 2013. http://ens-newswire.com/2013/02/page/2/

# 11

## THE ELEPHANT-IN-THE-ROOM

### CHINA – ASTOUNDING GROWTH AND STOLEN TIMBER

The astounding economic growth of China has attracted a host of superlatives. Its position as the largest importer of illegal timber is one of the undesirable ones. Although much of the wood processing industry in China is export oriented, a vast domestic construction effort, coupled with increasing wealth is creating a surge in demand for timber products. As we have seen, a vivid example is the fashion for furniture made from rare rosewood. Also a feature of the Chinese domestic scene since the late 1990s has been the strong measures taken to protect and grow its own forests.

To appreciate something of the current and likely future pressures on Southeast Asian jungles it is useful – if somewhat scary – to have a look at the emergence of Asia and the broad implications for manufacturing and consumption. A defining feature of the twenty first century is Asia's rise in global economic prominence that will have profound implications for people everywhere. Within a matter of years, Asia will not only be the world's largest producer of goods and services, it will also be the world's largest consumer of goods and services. China is already the most populous country in the world and in the future will also be home to the majority of the world's middle class.[145]

Asia's development continues against a backdrop of persistent global challenges. Among them is the weakness in major advanced economies, where only modest economic growth is anticipated. There are also global challenges like climate change and food, water and energy security. Also Asia has policy challenges of its own, including the continuation of market based reforms to promote economically and environmentally sustainable growth; dealing with urbanisation; an aging population, and international relations.

Over the past 20 years, not only in China, but right across Asia, the living

---

145 Based on: Australia in the Asian Century White Paper 2012, Commonwealth of Australia, Canberra.

standards for billions of people have improved markedly. In China and elsewhere in Asia technology and management systems have been imported and adapted further boosting productivity.

A more open global trading regime has enabled Asia's rising economies to more easily secure energy and resources critical to growth. This is particularly important for economies, such as China, Japan and South Korea that lacked large resource endowments themselves.

Asia's impact on global markets has been, and is likely to continue to be profound. The region's industrialisation and urbanisation has sparked demand for raw materials, creating a global resources boom. Asia's expanding middle class will see Asian economies emerge as the world's dominant consumer markets. This will reshape global markets, including for high value consumer goods and services – from electronics to tourism.

Despite their smaller size relative to China, Indonesia, Malaysia and other Southeast Asian countries are expected to grow rapidly and make solid contributions to regional growth. Along with China these countries will collectively alter nature of the global economy and balance of influence. Clearly Southeast Asia's cumulative economic and political weight is growing.

Overall, Asia will become more important in global trade to 2025, retaining its many advantages as a mass producer of manufactured goods. These trends will change spending patterns, social and cultural preferences, the use of technology and occupations throughout Asia and around the world.

As China's economy and status in the world further strengthens, you have to wonder when and where it will all end – one thing is for certain – the outcome for the jungles of Southeast Asia is likely not to be good.

Perhaps not the best analogy but certainly in terms of illegal logging China is the elephant-in-the-room. As China has become the world's biggest importer of timber products, it has emerged as the leading destination for illegally logged timber, especially logs. As illegal logging and timber smuggling are, by their very nature clandestine criminal activities, an exact quantification of the volume of stolen wood entering China is challenging. However, by combining wood flows with rates of illegal logging in source countries, credible estimates can be made.

China's huge and growing demand for 'raw' logs has dominated the country's timber imports throughout the past decade or more, with an average 33 per cent of its log consumption met by imports between 2000 and 2010. From imports in 2000 of 13.6 million cubic metres valued at $1.6 billion, China's log imports by 2011 had more than trebled in volume terms to 42 million cubic metres worth $8.2 billion. China imported around 30 per cent of all logs traded worldwide in 2011.

Russia was China's top log supplier in 2011, with exports of 14 million cubic metres; this despite a significant drop in Russian log shipments to China since 2007, partly as a consequence of an increase in log export taxes. Log demand in China is being felt around the world with the number of countries from which it sources logs is increasing.

The Environmental Investigation Agency's field investigations into flows of illicit timber into China indicated that illegal log imports alone have reached 11.8 million cubic metres, worth $2.7 billion.[146] Illegal logs made up 54 per cent of the volume and 58 per cent of the value of all log imports from the 36 countries surveyed. Major flows of illegal logs included 5.6 million cubic metres from Russia; 2.5 million cubic metres from PNG; 1.5 million cubic metres from the Solomon Islands; 500,000 cubic metres from each of Myanmar and the Congo; 270,000 cubic metres from Equatorial Guinea, and 183,000 cubic metres from Mozambique.

Environmental Investigation Agency research also reveals that Chinese state owned companies have imported logs from countries where such exports are banned, such as Indonesia and Mozambique, and are directly involved in logging operations in countries where illegal logging is prevalent.

While China's status as the world's biggest importer of illegal timber is in large measure due to its emergence as a major wood processing hub, and its burgeoning domestic market, it is also a consequence of the lack of strong policy direction by the government.

China's role as the world's biggest timber trader means that further progress against illegal logging depends on the nation taking measures to exclude illicit timber from its supply chain. Yet while other major consumer markets have acted, China remains somewhat on the sideline. To date, China's principal response has been to conclude a series of bilateral agreements with both producer and consumer countries. While these agreements have resulted in some progress, such as a drop in cross border log trade with Myanmar, generally they provide a mechanism for discussion rather than

146 *Appetite for destruction – China's trade in illegal timber* 2012 Environmental Investigation Agency, London, UK.

action, and have failed to constrain China's imports of illegal timber.

To date a major barrier to the effectiveness of China's anti illegal logging response has been the country's stated unwillingness to explicitly prohibit illegal timber trade. Although China's State Forestry Administration has engaged with the international community on illegal logging, it has resisted calls for legislation prohibiting illegal trade into and within the country.

# 12

## INCREASING THE OIL PRESSURE

### PALM OIL – DRIVER OF DESTRUCTION

We have spoken about illegal logging activities – bad enough you might think – but nothing like the destruction wrought on Southeast Asian jungles as a result of the rush to plant palm oil.

Palm oil production is clearly head-and-shoulders the major 'driver' of tropical forest destruction. Reporting in the *Washington Post* on 26 November 2012, Jason Motlagh[147] described how palm oil is now in great demand in western markets because of its low price and long shelf life. Derived from the fruit of oil palm trees, it can be found in many of products sold in western supermarkets, from cookies to cosmetics. Its use is continuing to increase as the commercial food industry phases out trans fats.[148]

Unfortunately the effects of Western consumer behaviour can have distant ecological and social impacts without consumers being conscious of this reality. The oil that almost everyone now uses, but few know, is obtained from the plum sized fruit of the African oil palm.[149] With a production of about 30 million tonnes in 2005 palm oil became the world's most important plant oil overtaking soy oil.

---

147 Jason Motlagh is a writer, photographer and filmmaker. He has served as *TIME Magazine's* Kabul correspondent, he has reported from around Afghanistan and more than 35 countries for leading international media, including *The Economist, Washington Post, New Republic, US News & World Report* and *Frontline/WORLD*.

148 There are two types of trans fats. The first occurs naturally in dairy products and accounts for about three quarters of the trans fat consumed in western countries. The second and dangerous form of trans fat was largely introduced into the diet in the 1950s with the rise of processed food and, in particular, margarines. Manufacturers found out how to make liquid oils remain solid at room temperature extending their shelf life and creating trans fat. This type of trans fat presents an increased health risks, increasing the bad type of cholesterol and elevating the risk of heart disease.

Trans fats can not only change the levels of cholesterol, but can cause inflammation in the body that contributes to many metabolic diseases, such as diabetes. The World Health Organisation recommends that less than one per cent of the diet is made up of trans fat.

149 African oil palm (*Elaeis guineensis*)

*Palm oil plantation: a major 'driver' of tropical forest destruction (JJ_Ch12_1).*

No other domesticated plant produces as much oil per hectare as the fast growing 20 metre tall palm oil providing its first harvest just three years after planting. The fruit bunches of palm oil contain several thousand fruits that are cut off when ripe broken up then steam sterilized to render an enzyme harmless which would otherwise make the palm oil rancid. The oil is pressed from the pulp and clarified.

Palm oil – coloured orange-red by carotenes[150] – is solid at room temperature and tastes sweet. From the mechanically broken up kernels expensive palm kernel oil is produced. The residue of the pressing, called palm kernel grist, is sold as animal feed.

Palm oil is an ingredient in thousands of everyday products. It is found in frying and cooking oils as well as in margarine, mayonnaise, ice cream, chocolates and sweets, soups, flavourings, instant meals, bread and cakes, cosmetics, soap and other personal hygiene items, in candles, detergents and cleaning agents – even in milling grease for the production of steel plate.

The presence of palm oil usually does not enter into the awareness of consumers as there is no explicit product labelling obligations. However, whenever a product is said to contain vegetable oil or fat, in all probability it contains palm oil. The huge global appetite for palm oil is now yielding

---

150 Carotenes are orange photosynthetic pigments important for photosynthesis. They are responsible for the orange color of the carrot and for the colors of many other fruits and vegetables. Carotenes are also responsible for the orange colours in dry foliage.

billions of dollars in revenue for Indonesia and Malaysia, the world's first and second largest producers of palm oil.

*Palm oil fruit: an ingredient in thousands of everyday products (JJ_Ch12_2).*

It is clear that this boom is doing irreparable damage to jungle biodiversity and also having related detrimental social consequences. A further adverse effect of the palm oil boom is that tribal people are being pushed off their ancestral land to make way for plantations staffed mostly by migrant workers. These imported workers are often denied basic health care, education services and any legal status.

Jungle cover in Borneo started to be lost to palm oil in the 1960s when the Malaysian government pushed the expansion of palm oil plantations to supplement its rubber tree industry. Indonesian and the Philippines migrant workers arrived in large numbers to work on the plantations established in cleared jungle.

According to the seminal book: *Thinkers of the Jungle*[151] orangutans have been the big losers in the race to establish extensive tracts of palm oil plantations. This is one of many chilling accounts from the book:

> The plantations in which the monumental stumps of the felled jungle giants frequently still project out of the ground are green deserts in which no bird flies and no lizard creeps. Quite often half starved orangutans – often fleeing from unstoppable forward-moving wood-cutter gangs – lose their way in the oil plantations where nothing edible exists for them apart from the marrow of palm trunks and get gunned down as vermin when they tear the trees apart.

*Dying orangutan: concern about the impact of palm oil plantations on orangutan survival is gathering momentum.*

No doubt about it, palm oil is bad news for Southeast Asian jungles. There is a gold rush going on and the name of the gold is palm oil. In Sumatra for instance there exists the prospect that another one and a half million hectares of jungle will be levelled and a million more in Borneo to establish new palm oil plantations.

Western consumer countries are helping to finance this ecological

---

151 *Thinkers of the Jungle* 2008, Gerd Schuster, Willie Smiths and Jay Ullal HF Ullman (www. ullmann-publishing.com) Germany.

holocaust – that in the case of orangutan – has seen vast areas of habitat and thousands of orangutans lost. Conservation director of the Sumatra Orangutan Conservation Programme,[152] Ian Singleton says the loss of orangutans has never been greater than in recent years and can be laid at the feet of the palm oil industry. "The problem is really an immense one."

Growth in Malaysian palm oil plantations has seen the industry evolve into that country's most lucrative crop. In 2011, the export of palm oil and palm based products netted $US27 billion – a fivefold increase over the past decade – thanks to substantially increased trade with China, Pakistan, Europe, India and the USA.

Indonesian officials have announced plans to convert about 18 million more hectares into palm oil plantations by 2020. Malaysia wants to double the area under cultivation over the same period to drive development in eastern provinces, where infrastructure and living standards lag behind the wealthier, more industrialised western peninsula.

Concerns about the impact of palm oil plantations on the Southeast Asian jungle habitat of the orangutan and other threatened Southeast Asian signature animals, and on the longer term survival of the jungle itself is gathering momentum. Supermarket chains in Europe, USA and Australia are feeling the heat as consumers question the fate of iconic animals and the palm oil lead destruction of their jungle homes.

A joint study published by Stanford and Yale Universities[153] shows that deforestation for the development of oil palm plantations in Borneo is becoming a globally significant source of carbon dioxide emissions. Plantation expansion is projected to contribute more than 560 million tonnes of carbon dioxide into the atmosphere in 2020 – an amount greater than all of Canada's current fossil fuel emissions. In 2010 alone land clearing for oil palm plantations in Kalimantan emitted more than 140 million tonnes of carbon dioxide – an amount equivalent to annual emissions from about 28 million vehicles.

---

152 The Sumatran Orangutan Conservation Programme is a collaborative programme of non government organisations and the Indonesian government working for the survival of the orangutan in Sumatra.
153 *Carbon emissions from forest conversion by Kalimantan oil palm plantations* 2013 Kimberly M. Carlson, Lisa M. Curran,Gregory P. Asner, Alice McDonald Pittman, Simon N. Trigg and J. Marion Adeney, Nature Climate Change Vol 3 283–287

*Another orangutan fatality, Sumatra: palm oil plantation expansion is bad news for orangutans*

Professor of ecological anthropology at Stanford University Lisa Curran said:

> Despite contentious debate over the types and uses of lands slated for oil palm plantations, the sector has grown rapidly over the past 20 years. These plantation leases are an unprecedented 'grand scale experiment' replacing forests with exotic palm monocultures. We may see tipping points in forest conversion where critical biophysical functions are disrupted, leaving the region increasingly vulnerable to droughts, fires and floods.

Commenting on the study Laurel Sutherlin of the San Francisco based environmental organisation Rainforest Action Network said:

> It's a perfect storm of human rights abuses and social conflict on the one hand and the destruction of some of the most biologically diverse forests in the world on the other.

According to an analysis by Reuters, major oil palm companies operating in Indonesia have been expanding at an average rate of 10,000 hectares a year per company.[154] Carbon dioxide emissions from the conversion of peat lands to palm oil plantations could be as high as 60 tonnes per hectare per year. Peat lands will continue to emit greenhouse gases even after land conversion activities have been completed.

In response to this mounting pressure, leading palm oil producers have partnered with advocacy groups to form the Roundtable on Sustainable Palm Oil,[155] an association based in Zurich aiming to establish clear social and environmental safeguards for the industry. Top consumer goods companies, such as Unilever and Nestle are members, as well as agribusiness giant Cargill, the largest importer of palm oil to the USA. Activists remain sceptical saying there has been more talk than serious reform and that a lot more needs to be done.

---

154 *Land banks buffer Indonesian palm oil from forest ban* 2011 Koswanage, N. And Taylor, M Reuters (25 May 2011) www.reuters.com/article/2011/05/25/us-indonesia-palmoil-forests-idUSTRE74o2LA20110525

155 On 8 April 2004, the Roundtable on Sustainable Palm Oil was established under Article 60 of the Swiss Civil Code with a governance structure that ensures fair representation of all stakeholders throughout the entire palm oil supply chain. The seat of the association is in Zurich, Switzerland, the Secretariat is based in Kuala Lumpur with a liaison office in Jakarta.

The inaugural meeting of the Roundtable took place in Kuala Lumpur, Malaysia on 21-22 August 2003 and was attended by 200 participants from 16 countries. The key output from this meeting was the adoption of the Statement of Intent which is a non legally binding expression of support for the Roundtable process.

# 13

## CONFRONTING THE TWIN EVILS

### ILLEGAL LOGGING AND CLEARANCE – THREATS AND SOLUTIONS

Driven by the twin evils of illegal logging and clearance, the destruction of Southeast Asian jungle is having a catastrophic impact on a wide range of biological and social values. So what then is the answer to this pincer movement threatening the very existence of the region's jungles?

You might think that the fight for jungle survival is complex, involving population pressures; consumption demands of Western countries and from China; lack of effective government control; weakness and corruption, and so on – and so it is. But cutting to the quick – so to speak – the answer is really quite simple. It is to make jungle trees too valuable to chop down – making the jungle too valuable to remove. It is about making the prospective income from keeping trees standing and the jungle intact outweigh the short term commercial benefits from ripping out trees and converting the land to some other form of use.

While we have seen that illegal and poorly practiced logging plays a major role in jungle destruction, it may not be logging activities themselves that cause jungle removal, but subsequent land clearing. No doubt that clearance is facilitated when logging roads and poor management provide access for illegal loggers, small scale migrant farmers or larger scale palm oil developers.

It is clear that the greatest richness of plant and animal species across the tropics is concentrated in lowland jungles. So it is of particular concern that the once extensive pristine lowland jungles of the Philippines, Thailand, Vietnam, Sumatra and Borneo – rich in plant and animal species – have already been reduced to less that 10 per cent of their original area.[156] Yet jungle loss and degradation continues.

In Southeast Asian countries agriculture and forestry policies frequently

---

156 *Protected areas systems review in the Indo-Malayan Realm,* 1997 MacKinnon, J. IUCN, Gland, Switzerland.

encourage the further 'opening up' of remote jungle areas and the construction of new roads that lead to further illegal logging and encroachment.

It is apparent that a variety of government policies are promoting the acceleration of the rate of jungle loss. These policies include initiatives that encourage resettlement of 'undeveloped' regions; transport and communication policies that involve road building through untouched jungle; energy policies that promote the flooding of lowland valleys for hydroelectric power schemes; pricing policies for timber and agricultural products; financial incentives for agricultural plantations; the development of wood and pulp processing ventures, and land tenure change incentives that foster the settlement of 'frontier' areas by rural poor and migrants.

*Jungle 'development', Indonesia: government policies promoted the acceleration of clearing*

Yes I did say that the answer to jungle survival was simple – perhaps I should have said *sounds simple* – as the policies and practices needed to promote jungle survival are demanding and present Southeast Asian countries with substantial challenges.

No doubt we want to protect the biological wonders of the jungle; the contribution plants and animals make to the diversity and quality of life, and the critical role jungles play in supporting the welfare of dependent tribal communities and rural poor. More broadly we want to protect the present and potentially still greater contribution jungles can make to the future economic fortune of Southeast Asian countries, and indeed countries across the globe.

Increasing pressure on jungles and on other natural resources reinforces the importance of political commitment and good governance to ensure sound jungle management. The greatest political challenge is to address the root causes of biodiversity loss and the perverse economic incentives that fail to recognise biodiversity conservation as a key to long term sustainable economic and social development.

Policy makers, government officials, the scientific community, conservationists industry companies, non government organisations and local communities need to work together to ensure that biodiversity options and concerns are integrated into government policies and development programs.

The economic argument to save the jungles must be framed in terms that promote *vertical* trees as being more valuable than *horizontal* ones. That keeping trees standing is a commercially compelling option. That maintaining jungle and its habitat values is economically worthwhile – even better that oil palm or agriculture. But how can this be done? The answer lies at least in a couple of obvious areas and in others less so.

In response to endeavours to tackle climate change the emergence of carbon dioxide emissions trading schemes provide the opportunity to generate carbon 'offsets' under cap-and-trade emissions trading schemes[157] being

---

157 Cap-and-trade emissions trading is a market based approach used to influence the adverse impacts of climate change by providing economic incentives for achieving reductions in the emissions of greenhouse gases.

Generally such emission trading schemes involve a government entity setting a limit, or *cap*, on the amount of greenhouse gases that may be emitted. The limit or cap is allocated or sold to companies in the form of emissions permits which represent the right to emit or discharge specific volumes of greenhouse gases. Companies are required to hold a permit (or allowances) or carbon offset credits equivalent to their emissions. Companies that need to increase the volume of their emissions are required to purchase carbon offsets. The transfer of carbon offsets is referred to as a trade. In effect, the buyer is paying a charge for polluting, while the seller is being rewarded for having reduced emissions or for generating carbon offsets, from say preventing forests being cleared or by planting trees.

implemented in many countries. There is already plenty of detail available. For example, Reduced Emissions through Deforestation and Degradation (REDD) schemes provide a means of paying to keep trees standing by essentially selling the carbon stored in these trees – in trunks, even roots and in some cases the soil – as carbon offsets. The revenue regenerated from the sale of such carbon offsets can be spread across the supply chain – if you like – from national and local governments to companies with legal entitlements over the land right down to local communities. So we change the dynamic from the view that the jungles are economically 'worthless' to seeing their retention as commercially valuable.

Yet another strategy is to encourage sensitive, sustainable timber harvesting coupled with utilising timber produced in the manufacture of high value products, such as plywood, engineered structural products, architect inspired buildings, furniture and so on.

*Previously logged jungle: a need to encourage sensitive and sustainable timber harvesting*

The revenue generated from sophisticated, low impact, sustainable logging could assist in improving practices and meeting the costs of careful harvesting; sustainable forest management; third party certification of practices and production, and in assisting to meet local community aspirations. Again, all directed at keeping jungles functioning as efficient, healthy ecosystems that maintain biodiversity, habitat value and economic returns.

So we now understand that sustainable management and the retention of a 'permanent' jungle estate is important both for biodiversity conservation and human welfare, especially, as we have said, for those communities most dependent on forests for subsistence and income.

In many cases non timber forest products, including animals and fish, have a substantial value to local and national economies that may exceed the value of trees as timber. Plant products from Southeast Asian jungles include a wide variety of food, fibre, fuel, beverages, horticultural plants, antioxidants, sources of chlorophyll, enzymes, food colourings, sweeteners, vitamins, medicines, fodder for domestic animals – the list goes on.

So collectively carbon offsets and careful harvesting of timber from sustainable, certified forest operations provide part of the 'formula' to reverse current trends of continuing the degradation and destruction of Southeast Asian jungle. Such a formula offers a pathway down which humanity *must* walk if we are going to save the jungles of Southeast Asia. What do you think?

In this chapter we have talked about a number of actions that might individually and collectively assist in providing a more optimistic future for the jungles of Southeast Asia, together with their human and other treasured inhabitants. We will explore ideas related to sensitive, sustainable management; enhanced timber use; climate change abatement and carbon offsets, and more sympathetic policy settlings that have jungle conservation as a key cornerstone in the chapters ahead.

# 14

## Nature's Perpetual Motion Machine

### Sustaining Jungle Tree Growth

I have tried my hand at a few things work wise, but basically and proudly I am a forester.[158] I have spent a fair bit of my working life dealing with issues related to the management of natural forests – so you would think I should know a thing or two relevant to this chapter. Let's see. Here we will be dealing with the jungles of Southeast Asia – natural forest systems – as opposed to planted forest.

But before looking at forest management policies, practices and controls, let us attempted to get to grips with some of the definitions and principles that are central elements of this chapter, and the quest for the sustainable management of Southeast Asian jungle.

In the context of forest management, the first logical question to ask is what exactly it is that we are trying to sustain? Sustainability is a loosely used word – frequently thrown around without much quantification or clarification. I could be pedantic here and go on about ecological sustainability and what I think that means, but I won't. In relation to tropical forests – complex and difficult to manage as we will see – if we can sustain, or perpetuate if you like, the biomass[159] of the jungle, particularly of the canopy trees, we will have done exceedingly well. If we can achieve that other things will follow. So let us investigate this definition of sustainability.

158 Historically foresters were game keepers. Their main duties related to game keeping and preventing poaching on behalf of land owners. This changed in Europe in the 17th century when interest in sustainable timber production started to emerge and foresters assumed responsibilities for forest management and related technical disciplines.
159 In the context of this book biomass refers to the above ground plant material, notably the wood stored in the trunks of trees. Biomass may also be used to describe all of the living plant material in the jungle, including branches and leaves from trees, shrubs and other plants.

*Managed forest, Malaysia: aiming to sustain jungle biomass*

Maintaining the density of canopy trees within prescribed geographical boundaries and time periods will often mean that other jungle attributes, like biodiversity and habitat quality will also be maintained. Generally we might want to supplement forest management with the addition of practices, like not removing trees adjacent to streams, and leaving old and over mature trees that provide important habitat for birds and tree inhabiting animals. Putting these sorts of considerations to one side for a moment sustaining the biomass or *sustained yield management* – to use the forestry term – means maintaining the volume of trees over both time and space.

So felling and harvesting trees is fine as long as doing so does not damage the ability of the jungle to regenerate and grow new trees. In practical terms this means that over a defined area – that might be some thousands of hectares – the overall volume of the biomass remains relatively constant. Trees are felled and removed and new trees grow to replace them. The critical part of the equation is that the natural built in ability of the jungle to replace trees that fall naturally or are removed by logging is not compromised – we will come to this critical issue a bit later.

All too often in the past jungles have been over logged – too many trees have been removed from a given area and too much damage done. This results in the natural renewal mechanism of the jungle ecosystem breaking down. Invasion of 'weed' species, notably climbing plants that rapidly colonise openings in the canopy, smothering the ground and preventing tree species from regenerating is a common symptom of such over logging.

Over logging means the multi layered natural dynamics of the jungle are damaged. Remaining trees are exposed to the weather – storms and high winds – so toppling and further losses of canopy trees can occur. Because the original jungle structure has been compromised and degraded the normal systems of growth and renewal begin to fail. The quality of the jungle in terms of habitat value and biodiversity also diminishes.

On the other hand, sensibly managed jungle, based on sustained yield forestry practices where overall growth capacity is maintained means other jungle values – whilst temporarily impacted in a particular location – may be maintained over larger tracts.

Experience indicates that, at least in some cases, the increased plant diversity caused by tree removal may actually benefit some species. Ground cover plants temporarily become more abundant providing grazing for plant eating animals before regenerating tree species again dominate the site, crowding out ground cover species. I am going to return to a variation on this theme in relation to the potential benefits of sensitive logging in assisting with the maintenance of orangutan habit.

So a fundamental principle of forestry management in the tropics, and elsewhere, is to harvest no more wood than the forest can regrow. This sustained yield management means we live off the wood 'interest' from the wood 'capital'.

This was the cornerstone of acceptable forest management in Australia to which Antarctic explorer Sir Douglas Mawson referred in 1925 when he warned about the rate of forest exploitation and the depletion of timber resources. He drew an analogy with raiding the bank account of: "future generations of Australians" rather than living off the "interest" or annual yield accrued by the growing "capital stock" that was the forest.[160] Mawson's summary of sustainable forestry practice applies equally to the tropical jungles of Southeast Asia.

As one of the world's more famous explorers, Mawson made his name traversing an ice contingent – without a tree in sight! He led the 1912 Australasian Antarctic Expedition[161] that was at the time the most ambitious exploration and research expedition ever undertaken to Antarctic. Included in the list of science disciplines studied by the expedition were geology, meteorology, magnetism, biology, atmospheric science and glaciology. Mawson set off in November 1912 with two companions. He spent weeks in the open and struggled back to based camp in February 1913 barely recognisable and the sole survivor of the group.

An Australian hero Mawson was knighted by King George V and subsequently had a distinguished career as a professor at the University of Adelaide in South Australia where he lectured and led research in a number of scientific endeavours – including forest science.[162]

*Douglas Mawson: famed Antarctic explorer also lectured in forest science*

---

160 More details in Chapter 9 of *Trees that call Australia home* 2008 John Halkett Potts Point Publishing, Sydney, Australia.
161 A 31 strong science team.
162 For an illuminating summary of Mawson's Antarctic adventures see: *Into the unknown*, by David Roberts in the January 2013 edition of *National Geographic*.

Going back into history a bit before Mawson, it was not really until the middle of the 17th century in Western Europe that the first formal attempts at 'modern' forest management began. The concept of sustained yield management evolved in Germany in the years following the traumatic Thirty Years War.[163] During the period of stability that followed the war population increased, cities and towns were rebuilt and there was a rapid expansion of industry. It was soon realised that the depletion of forests was looming as a widespread and serious problem. This lead to severe timber shortages that were accompanied by flooding, erosion and landslides as forests in mountainous regions were cut to support booming residential building and industrial activities.

Forest conservation ordinances were introduced and trees planted, protected and tended in accordance with prepared working plans. The concept of forests supplying the timber needs of communities started to be documented. Foresters began to measure tree volumes, assess growth rates and timber yields, and to prescribe the annual allowable harvest of merchantable timber in accordance with the evolving concept of sustained yield timber management. These developments in Europe continued into the 18th and 19th centuries.

As a maritime trading nation and at the time possessor of an extensive empire of forest rich colonies Britain had no need for sophisticated forestry practices. It was not until the British Isles was blockaded by German submarines in the First World War that Britons realised there was a need to nurture and sustain their own forests. After the settlement of North America the vast natural forests were regarded as limitless, or worse as a liability to be used or otherwise destroyed to make room for settlers and agriculture.

---

163 The Thirty Years' War (1618–1648) was a series of wars principally fought in central Europe, involving most of the countries of Europe. It was one of the longest and most destructive conflicts in European history and one of the longest continuous wars in modern history. The origins of the conflict and goals of the participants was complex and no single cause can accurately be described as the main reason for the fighting. Initially it was fought largely as a religious war between Protestants and Catholics, although disputes over internal politics and the balance of power within the Empire played a significant part.

A major consequence of the Thirty Years' War was the devastation of entire regions by foraging armies. Famine, disease and forest destruction significantly decreased the population of the German states and Italy.

When I went to forestry school in the 20th century there were a couple of luminary figures that had an impact on modern forestry thinking and education – one German and the other an American. Modern forestry practice was still substantially influenced by practices in Europe and by the writings of German born forester Wilhelm Schlich, and later by the progressive thinking of well known American forester, Gifford Pinchot.

Born in February 1840 Wilhelm Schlich entered the British Imperial Indian Forest Service in 1866, becoming Conservator of Forests in 1871, and Inspector General of Forests in 1883. He developed forest management and education programs during the nineteen years he spent in India. He moved to England in 1885 and was one of the pioneers of the study of forestry organising the first specialist school at Cooper's Hill, Gloucestershire that was transferred to Oxford in 1905. He was appointed Professor of Forestry at Oxford the same year. He was made a Fellow of the Royal Society in 1901 and awarded the Knight Commander of the Indian Empire in 1909.

Schlich was the author of the five volume *Manual of Forestry* published from 1889 to 1896. Schlich's *Manual* became the standard and enduring text book for forestry students. He died in Oxford in 1925.

Schlich was a colleague and mentor of American forester and high profile politician Gifford Pinchot (1865-1946). Pinchot served as the first Chief of the United States Forest Service from 1905 to 1910 and was the 28th Governor of Pennsylvania serving from 1923 to 1927 and again from 1931 to 1935.

Pinchot is acknowledged for reforming forest management in the USA and for advocating the conservation of the country's forests by planned use and renewal. He coined the term *conservation ethic* as applied to natural resources. Pinchot's legacy was his leadership in promoting scientific forestry and emphasizing the controlled, profitable use

*Sir Wilhelm Schlich: pioneer in the study of forestry and author of the five volume 'Manual of Forestry'*

of forests and other natural resources. He was the first to demonstrate the practicality and profitability of managing forest for 'continuous cropping'. His leadership put the conservation and wise use of forests on the American priority list.

As we will see later, as well as an appreciation of botany, foresters need an understanding of tree population dynamics, statistical design and analysis, economics, engineering and environmental sciences.

All very interesting, but let us return to sustained yield forest management. As outlined, in its most simple form sustained yield management means balancing wood harvest and forest growth so that timber is produced indefinitely. However, like some other simple explanations, sustained yield management has frequently proven harder to practice than to preach. Although the principles can be readily explained they are less easy to implement, particularly in diverse and complex forest systems like the jungles of Southeast Asia.

In its basic arithmetic form sustain yield management means that annual timber harvest must equal annual growth, or expressed another way, the volume of timber that can be harvested each year, or from time-to-time, must equal the volume grown over that same period of time.

Returning to Mawson's banking analogy, if you say invest $1000 in the bank and it earns $50 interest each year you can spend $50 each year *forever* without affecting the value of the capital – assuming of course that the interest rate doesn't change, and you keep your hands off the capital.

So now imagine that a forest of a defined size managed for a sustained yield of timber is assessed as containing 50,000 cubic metres of merchantable logs.[164] The forest is also assessed as growing 500 cubic metres of merchantable logs every year. This increase is made up of the growth of young trees to a size big enough to qualify as being merchantable, plus the increased growth of existing merchantable trees, less any losses through the death of merchantable trees.

This means that so long as the forest area and fertility remain constant, and provided trees removed are regenerated so that the overall net growth of timber remains at 500 cubic metres of merchantable logs, then 500 cubic

164 Merchantable logs is a classification indicating that the logs are of dimensions – diameter, length, and form –that makes then suitable to be processed into the usable products for which they have been grown.

metres of merchantable logs can be harvested each year, essentially forever without deleting the forest's growing capital stock of trees.

*Reviewing forest management practices, Malaysia: aim is to continue harvesting without depleting the growing stock.*

Unfortunately life is never that simple, so a few cautionary notes. Firstly the sustained yield model is just that, an arithmetic model based on data related to a specific area of forest, its growth rate and the merchantable dimensions of logs harvested. The varying influence of the market place, coupled with economic and political factors are conveniently ignored in the model. However, in the real world we know that timber 'consumption' does not follow the rules of arithmetic but varies according the economic conditions and fluctuations in trade and commerce. It is also likely to be affected by changing building technology, fashion trends, population demographics and so on, and so on.

So we see that while the theoretical sustained yield model has appeal, it does not always fit the realities of the real world. Market and economic influences and other factors, including the possibility of fires or other natural disasters affecting the forest may dictate a need to 'overcut' for a period of time. Conversely, there may be times when the market doesn't want the timber available according to the sustained yield model or for whatever reason the forest fails to produce the predicted yield. In such circumstances an 'undercut' will be forced on forest managers.

As we have learnt from earlier chapters a look at any area of tropical forest will quickly tell you that, when compared to temperate forests, they appear as a jumble of vegetation, often intertwined and connected. Remember in Chapter 4 we said that while tropical forests have a definite structure they can appear as a bewildering chaos of vegetation. I quoted Friedrich Junghuhn who suggested jungles exhibited *horror vacui – fear of empty space* – where the jungle seems anxious to fill every available space with trees, branches, twigs and leaves. This is a good visualization of the tropical jungles of Southeast Asia, and points to some real challenges for the practical application of the sustained yield model.

So applying the principles of sustained yield management to complex jungle is a challenging issue. Traditional forestry practices used successfully in temperate forests have, when applied to tropical forest, often led to disappointing outcomes. Past logging operations in Southeast Asian jungles have frequently resulted in extensive roading and tracking; excessively large log storage areas; wide canopy openings resulting in 'invasion' by vines and 'weed' plant species, and substantial damage to residual canopy and sub canopy trees.

Because of these factors, techniques such as reduced impact logging have been developed to move tropical forest management more in the direction of the sustainable management model we have been discussing. Reduced impact logging[165] is essentially a set of techniques that offer a more sustainable approach to log harvesting in complex jungle environments.

Reduced impact logging techniques focus on tree harvesting at prescribed levels using practices that minimise 'collateral damage' to non harvested trees and to wildlife, streams and other natural values. Undertaken on a 25 to 50 year cycle, depending on forest type and its productivity, reduced

---

165 For more information about reduced impact logging and its application in tropical forests visit the Tropical Forest Foundation website: www.tropicalforestfoundation.org

impact logging is aimed at minimising disturbance so that log harvesting, while providing economic benefits, is also directed at retaining the values associated with unlogged areas. The challenge is to produce an economic benefit while at the same time maintaining the sustained yield capacity of the jungle and protecting other values.

Reduced impact logging covers the full spectrum of forest harvesting operations, from pre logging inventory and planning; selection of merchantable trees; design of roads and tracking; tree felling, extraction and hauling of logs, and finally post logging operations and assessments. Where it has been practiced in Southeast Asia it has been shown to be an important step towards responsible and sustainable management, demonstrating an ability to reduce soil disturbance in roading, log storage areas and machinery tracking by almost half; significantly reducing large canopy openings resulting in better survival and faster recovery of residual trees; lower log waste, and minimising machinery operating times.

Well so much for all the theory I hear you say, but how does sustained yield management actually work in practice? Let us look at an example – take say 1500 hectares of a random jungle logging concession somewhere in Southeast Asia. The first step is to undertake some planning guided by the relevant forestry code of practice. Typically a code of practice contains prescriptions about operational details related to log harvesting, such as tree removal levels; machinery tracking; road construction and soil disturbance, water and environmental protection.

Using the prescriptions in a code of practice, plus inspections and assessments in our 1500 hectare example it is determined that 500 hectares will not be logged because of a combination of factors. These include that the area is either too steep or swampy for logging machinery to operate; the estimated merchantable volume of trees is insufficient to warrant the cost of tree harvesting; areas adjacent to streams needs to be excluded as waterway 'buffer' protection; some localised areas are recognised as particularly important for wildlife. The list of reasons for excluding specific areas might be longer.

Now that leaves us with 1000 hectares to log. You will notice that I am trying to keep the arithmetic simple. Data from a pre logging assessment – carried out possibly using sample plots or sample strips – has determined that on average the forest contains 600 cubic metres of biomass per hectare in the upper canopy of dominant trees. However, in terms of merchantable volume – using merchantability criteria related to tree size and form – plus selecting only those species recognised as having commercial value, it is determined that on average the merchantable volume per hectare is 250 cubic metres.

So we now know the available merchantable volume per hectare and will know from the code of practice the period between log harvests – or in forestry language – the *cutting cycle*. In this example it has been set at 30 years.

We should also know the growth rate of the dominant canopy forming trees from research and growth assessments. Let us say that the growth rate is ten cubic metres of tree trunk volume per hectare each year. However, in relation to log harvesting we are only interested in merchant trees, so with a bit of simple arithmetic that 250 cubic metres of merchantable tree volume will be growing at a bit more than four cubic metres per hectare each year.

Again a bit of arithmetic tells us that over a cutting cycle of 30 years, based on the tree population, growth data and cutting cycle length, the merchantable proportion of the area will be the four cubic metres growth each year multiplied by the cutting cycle of 30 years. That is 120 cubic metres of merchantable volume per hectare can be removed in logging operations conducted at 30 year intervals. It is also necessary to understand the average volume of trees deemed to be merchantable. Let's say that in this example it is at least three cubic metres of logs per tree.

If we also make a precautionary allowance to compensate for the fact that even using reduced impact logging techniques, particularly directional felling of trees and restricted machinery movement, some unintended damage and potential loss of growth will occur to future merchantable trees. From experience we assess this can be provided for by leaving 20 per cent of the volume of harvestable trees behind as an 'allowance' for this unintended damage.

So getting to the end of the arithmetic, we can now determine that for this particular cutting cycle we can plan and design the log harvesting operation to remove around 30 merchantable trees per hectare. It is actually 32 – you do the maths.[166] So in this example over the 1000 hectares demarcated for log harvesting we will produce about 90,000 cubic metres of merchantable logs.

---

166 120 cubic metres of merchantable volume per hectare, less 20 per cent allowance for damage; divided by the average size of merchantable trees assessed at three cubic metres of logs.

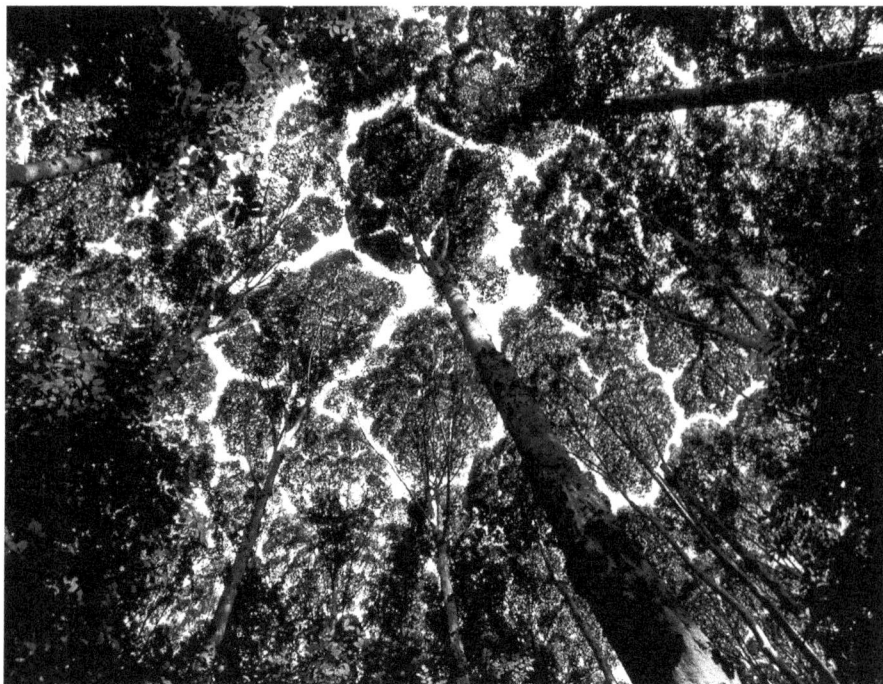

*Jungle canopy: need to know the growth rate of canopy trees from assessment and research work*

Before we wrap up this chapter let us return to the potential benefits of sustainable logging in assisting to maintain orangutan habit and populations in Borneo that I mentioned at the start of the chapter. A *PLOS ONE* online scientific article[167] analyzing an accumulated body of evidence indicated that selectively logged jungle in Borneo can play an important role in the conservation of orangutan populations.[168]

As a result, there is now recognition and increased understanding that logging concessions are an important component of maintaining jungle habitat for orangutans while at the same time promoting economic development. However, the articles point out that it is important to

---

167   *Understanding the Impacts of Land-Use Policies on a Threatened Species: Is There a Future for the Bornean Orang-utan?*, 2012 Wich SA, Gaveau D, Abram N, Ancrenaz M, Baccini A, et al. Ed: Sharon Gursky-Doyen, Texas A & M University, USA, PLOS ONE 7(11): e49142. doi:10.1371/journal.pone.0049142

168   Also see: *Orangutan distribution, density, abundance and impacts of disturbance.* 2009 Husson SJ, Wich SA, Marshall AJ, Dennis RA, Ancrenaz M, et al. In: Wich SA, Atmoko SU, Setia TM, van Schaik CP, editors. Orangutans: geographic variation in behavioral ecology and conservation. Oxford, UK: Oxford University Press. 77–96

recognize that sensitive management is crucial to the effectiveness of these forest concessions for orangutan conservation. Concessions where timber is harvesting has taken place at unsustainable levels tend to have lower orangutan densities.[169] Also logging concessions designated for sustainable, selective logging must be remained as permanent jungle.

The research cited contrasts this outcome with the reality that over past decades many logging concessions in Indonesia and Malaysia having been depleted of commercial timber because of over logging have been reclassified and cleared for palm oil plantation development or industrial tree plantations.

Selectively logged jungle areas constitute the highest percentage of the orangutan distribution of any land type in Borneo. Therefore the future survival prospects of the species depends largely on the strength of commitments by governments and companies to reduce deforestation and forest degradation rates in logging concessions and to maintain these forests for sustainable timber harvest over the long term.

Reaffirming this position Centre for International Forest Research researcher David Gaveau said that nearly a third of all orangutans left in the wild could be found within logging concessions in Indonesia:[170]

> With almost a third of remaining orangutans now living in forest concessions where the selective harvesting of trees for timber is permitted, we need to consider areas of sustainable logging as one way to protect the forest and orangutan populations.
>
> People need timber, and if you can produce it and at the same time protect wildlife, then we can actually achieve sustainable development.

Dr Gaveau argued that a number of factors threatened the long term future of orangutans, including deforestation for palm oil plantation development. He warned that about a quarter of remaining habitat was in jungle areas identified for development. Such areas comprised 19 per cent designated for palm oil plantations and 6 percent for tree plantation establishment.

In confirming that logging concessions were a viable option for protecting jungles, Dr Gaveau said:

---

169  *Aerial surveys give new estimates for orangutans in Sabah, Malaysia.* 2005 Ancrenaz M, Gimenez O, Ambu L, Ancrenaz K, Andau P, et al. PLOS Biology 3: 30–37. doi: 10.1371/journal.pbio.0030003
170  *Logging concession areas: Good for orangutans and forest conservation, says study.* In *Forests news* Center for International forestry Research, Bogor, BOGOR, Indonesia January, 2013.

Logging concessions can do just as well as protected areas in conserving wildlife, provided they stay as logging concessions and are not changed into zones that permit clearing for agricultural development.

This is crucial because governments in Indonesia and other tropical countries tend to reclassify logged areas for use in agriculture or oil palm plantations. So encouraging selective logging with rehabilitation and restoration, and discouraging conversion of logged forest to plantations could play a big role in helping protect orangutans.

I will draw this chapter to a close here. In the context of sustained yield management – and the pursuit of sustainable forestry practices – we need to return to some of the issues discussed in Chapter 10 related to illegal logging and in the next chapter talk about legality verification of timber products and forest certification schemes.

# 15

## YOU SHOULD BE CERTIFIED

### SUSTAINABILITY AND LEGALITY ASSURANCE

I am somewhat nervous in approaching this chapter. Any discussion about aspects of forest and timber product certification can quickly get bogged down in heaps of acronyms and technical shades of grey. To the extent that I can I want to avoid going too far down a quite complex technical path, when really the objective of forest and timber product certification is fundamentally straightforward. As discussed in the last chapter it is about ensuring that logging is undertaken in a sustainable manner, or at the very least, that timber products are made from logs that are sourced from legal forest operations. So it is about sustainable forestry practices or stopping illegal logging activity, or both. However, as we have seen in the context of Southeast Asia, these are substantial challenge. Challenges that also impact on the formulation, implementation and auditing of certification requirements.

When discussing illegal logging in Chapter 10 it was mentioned that it thrives in countries with poor forest governance and where law enforcement is weak. With these problems more associated with developing countries such nations experience the worst impacts of illegal logging, undermining longer term moves towards sustainable forest management.

As has already been stated illegal logging robs governments of income from log royalties and taxes. It also denies traditional owners income from their ancestral assets, and irreparably degrades forests, especially when protected areas are illegally logged. It has already been concluded that illegal logging and trade in stolen timber are among the most destructive of environmental crimes, threatening vital jungle ecosystems and often acting as a precursor to wholesale clearance. It has been noted that the loss of biodiversity and habitat of species, like orangutans, tigers and elephants resulting from illegal logging places such species even more at risk of extinction.

It was pointed out that violence and murder are often associated with the illegal timber trade and that in some instances financial flows generated by illegal logging have in some countries exacerbated armed conflicts.

It was also noted that in consumer countries timber products linked to illegal logging exert downward pressure on prices of legitimate products. Further, that the hint of illegal logging runs the risk of severely denting the outstanding climate change abatement, low energy manufacture and renewability credentials of timber products.

You will see as we get into this chapter, that there is a whole parade of forest and timber product certification and legality verification systems that have, or are being developed. We will also see that because determining sustainability and ensuring timber products are derived from forests so managed, or at the very least are produced from legal forest operations are all challenging tasks, certification systems are complex to design, implement and monitor.

*Checking log loading documentation, Malaysia: there is a whole parade of forest and timber product certification and legality verification systems that have, or are being developed*

In this chapter I will have a go at explaining some of the fundamentals of forest certification and legality verification systems – their benefits, operational details, remaining challenges and so on. For the sake of clarity I will be taking certification to mean certification in relation to sustainable forest management and associated timber product supply chain monitoring. When talking about legality verification, I will be referring to systems that relate solely to the legal status of logging activity. Such systems do not imply or assume that related forest operations are necessarily sustainable. I hope that is reasonably clear.

The consequences of weak forest governance and related poor practices in parts of Southeast Asia mean that many logging operations are far removed from those required by certification standards. This is the reality even though the legislative framework across the region is frequently considered to be adequate to achieve legal and even sustainable management status. Problems appear not to lie so much with the laws and regulations themselves, but with the failure or inability to implement them. Inadequate enforcement of forest regulations, combined with slow uptake and support for certification from logging companies and government agencies pose significant hurdles to the progress of legality verification and certification systems.

Looking at consumer nations, a leading catalyst north of the equator for implementing timber legality assurance requirements was the decision of the United Kingdom Government not to allow timber products to be purchased for government assisted projects where legality was an issue. This translated to about a third of timber product sales, so the industry, lead by the United Kingdom Timber Trade Federation, sat up and took serious notice. This initiative was followed by the large home improvement and trade hardware chains also starting to demand proof of legality for the timber products they handled.

These actions in the Northern Hemisphere got the attention of major timber product supply countries, including some in Southeast Asia. These countries, like Malaysia, accelerated efforts to put in place forest and timber product certification systems that they hoped would satisfy the policy and legislative aspirations of influential buying countries in Europe and the USA.

A key global response to forest management sustainability issues has been to develop third party forest and timber products supply chain certification systems. Amongst the largest are the Forest Stewardship Council (FSC) scheme established in 1993 and the Programme for the Endorsement of Forest Certification (PEFC) scheme established in 1999.

However, to date the majority of forest certified under these and other schemes are temperate forest and plantations. Further, the 'take up' of 'chain-of-custody'[171] certification by timber producers has been modest, with most of the interest coming from softwood timber product producers. In Europe and Canada it is estimated that less than five per cent of timber products in the supply chain are certified; in the USA less than two per cent, and in Japan less than half a per cent. Certified tropical timber products are available in only small quantities and from a less stable supply sources.[172]

Perhaps due to complexities, constraints, challenges and costs the attainment of certification has been at best marginal and tropical forests continue to lag well behind temperate forests in terms of area and products certified. However, this situation is showing some signs of improvement following commitments from governments and major companies to switch to certified suppliers; a growing consumer awareness about the impact of their buying choices, and new laws requiring legal and sustainable supply chains.

Turning to some background to legality verification, the USA became the first country to ban the import, sale or trade of illegally logged timber products by amending the *Lacey Act* in 2008. There are now strong penalties in the USA for knowingly sourcing, or failing to exercise 'due care' when buying what may turn out to be illegally logged timber.

The *Lacey Act* is actually a 1900 law enacted to ban trafficking in illegal wildlife. The original Act was directed more at the preservation of game and wild birds by making it a crime to poach game in one state with the purpose of selling it in another. It was also concerned with the potential problems of introducing exotic species of birds and animals into native American ecosystems. In 2008 the Act was amended to include plants and plant products, such as timber and paper. This landmark amendment to

---

171 Chain-of-custody is an inventory control mechanism that tracks a timber product from source in a certified forest through to end use. It covers all intermediate steps, such as logging, transportation, processing, manufacturing, distribution and sales.

172 *Global Forest Footprints: How businesses around the world contribute to deforestation – the risks of inaction and the opportunity for change* 2009 Mardas, N; Mitchell, A; Crosbie, L; Ripley, S; Howard, R; Elia, C, and Trivedi, M Forest Footprint Disclosure Project, Global Canopy Programme, Oxford, UK.

the *Lacey Act* amounted to the world's first ban on trade in illegally sourced timber products.

There are two major components of the *Lacey Act* amendment – a ban on trading plants or plant products, including wood harvested in violation of the law, and a requirement to declare the scientific name, value, quantity, and country of log harvest origin of imported timber products.

Andrea Johnson of the Environmental Investigation Agency in Washington DC summed up the implementation of the *Lacey Act* as follows:[173]

> With the new Lacey Act amendments, the landscape has changed for companies doing business with the US. Any business that wants to comply with Lacey, and avoid the risk of forfeitures, fines or even jail time, will be looking for timber sourced with careful attention to legality. Meaningful mechanisms for establishing and verifying legality, especially in high risk regions, will be critical tools to guide decisions about timber sourcing to minimise risk.

Across in Europe the European Union Timber Regulation was formulated to address illegal logging. This regulation came into effect on 3 March 2013. Under this regulation the European Union prohibits timber importers placing products derived from illegal timber on European markets and makes it a criminal offence to do so.

The European regulation requires importers to exercise 'due diligence'. Due diligence means European Union importers must minimise the risk of illegal timber entering the supply chain. In order to do this they need to have access to the information prescribed by the regulation, including the tree species, origin of timber products and compliance with national laws of the country where the logs from which the timber product was manufactured were harvested.

The European regulation covers a wide range of timber products, including fuel wood, plywood, timber, pulp and paper, furniture, joinery and other 'complex' products. It does not cover musical instruments and printed materials, such as books and newspapers.

Only timber products carrying Forest Law Enforcement, Governance and Trade or FLEGT licences are considered legal and to conform with the European Union Timber Regulation. In the absence of such a licence responsibility rests with timber importers to exercise due diligence by way

---

173 This quote from Andrea Johnson is taken from: *What Is a Voluntary Partnership Agreement? – the European Union Approach* EFI Policy Brief 3 2009 EU FLEGT, European Forest Institute.

of a risk management analysis based on relevant country and supply chain information, plus the criteria set out in the regulation.

In Australia the *Illegal Logging Prohibition Act 2012* is similar legislation to that in Europe and the USA. Australian legislation is intended to support the trade of legal timber into Australia and to restrict illegally produced timber from entering the market. The Act makes it a criminal offence to import timber products containing illegally sourced timber. Australian importers who "knowingly, intentionally or recklessly" import illegally logged timber products may be prosecuted. Criminal sanctions range up to five years' imprisonment and substantial fines.

The regulations to the Australian legislation that came into effect in November 2014 set out a due diligence process, contain a detailed list of products covered and an operational framework for importers.

An essential requirement of all legality verification and certification systems is that they have the capacity to track logs and timber products from the forest to the point of export, for the purpose of guaranteeing that exported timber products are derived from either legal or sustainable sources, and that timber derived from unknown or illegal sources is excluded from consumer country markets.

An effective system for ensuring timber product legality requires checking of forest operations, as well as transport and wood processing from logging to the point of export. A robust legality assurance system also needs a clear definition of legal timber; a mechanism to monitor timber products as they move along processing and supply chains; a government endorsed institution to determine compliance with relevant laws and possibly an independent institution to monitor the functioning of the whole system.

While current certification and legality assurance systems generally rely on conventional documentation, inspection and audit techniques, technology is increasingly assisting the process. For instance, bar coding and scanning devices are now used and an emerging technology in legality verification is the use of DNA[174] tracking relying on the genetic data inherent in wood DNA as a 'natural barcode'.

---

174 DNA or Deoxyribonucleic acid is a molecule that encodes the genetic instructions used in the development and functioning of all known living plant and animal organisms. Within cells, DNA is organized into long structures called chromosomes. During cell division these chromosomes are duplicated in the process of DNA replication, providing each cell its own complete set of chromosomes.

DNA tracking is innovative, cutting edge technology with the potential to enhance the verification of the source of timber, similar to its routine use in human criminal forensic science. DNA methodology is based on matching samples taken from the same log at different stages along the supply chain. A DNA fingerprinting test is used to confirm whether the samples originate from the same log, validating or invalidating chain-of-custody documentation. DNA methods have long been considered as a potential timber verification solution, having been discussed as part of a joint World Bank and World Wildlife Fund report on technologies for wood tracking in 2002.[175] A comprehensive report on DNA timber tracking commissioned by the Danish Ministry of the Environment in 2008[176] concluded that DNA analysis offered an exciting potential to establish the origin of timber products, but that significant challenges in its use remain. These challenges include the creation of a DNA 'geno-graphic'[177] database with enough genetic detail to determine timber product origin at a suitable scale and refining techniques of DNA extraction and analysis from finished timber products to facilitate a match against a database.

However, DNA tracking based on the matching of samples from a single tree does not require an existing genetic database. This concept is similar to a human paternity test, in which DNA samples taken from two individuals are tested against each other to see if they match.

DNA testing is not designed to replace existing paper based systems; rather it is designed to support, simplify and strengthen them. Genetic mismatches highlighted by DNA testing can act as a 'red flag' to auditors, who can then conduct more thorough investigations. By relying more on scientific evidence, the frequency and intensity of physical chain-of-custody audits can be reduced, bringing down the overall cost of certification. Reducing the cost makes it more viable for adoption of chain-of-custody certification schemes by a greater number of timber product manufacturers and exporters.

---

175 Technologies for Wood Tracking: Verifying and Monitoring the Chain of Custody and Legal Compliance in the Timber Industry, 2002 Dykstra, D.P., Kuru, G., Taylor, R., Nussbaum, R., Magrath, W.B., and Story, J., World Bank.
176 *Tracing Timber from Forest to Consumer with DNA Markers*, 2008 Nielsen, L.R. and Kjaer, E.D. Danish Ministry of the Environment, Forest and Nature Agency.
177 A geno-graphic database is created by analysing specific genetic markers in the DNA that are unique to a particular tree species, but vary across geographic regions. The development process involves DNA testing of hundreds of wood samples and mapping genetic variations. Genetic markers can be used for various applications in the study of genetics including identification of populations.

In relation to certification and legality verification more broadly, the heart of the concept of due diligence is that timber importers in the USA, Europe and Australia must undertake a risk assessment to minimise the chance of placing products containing illegally logged timber on the market.

Basically there are three elements of any due diligence system. The first is the gathering of information related to the description the timber products in question, their country of harvest, details of the supplier and information on compliance with relevant legislation of the country where the timber product was logged. The second element relates to the need for the importer, based on the information obtained, to assess the risk of illegal timber entering in the supply chain.

Finally, if and when the assessment shows that there is a risk of illegal timber entering the supply chain that risk needs to be mitigated by seeking additional information and verification from the product supplier to more precisely determine the extent of the risk. A determination that the risk is too high would mean that a particular timber product should not be imported.

Although the onus in meeting certification and legal verification requirements rests with importers, exporters of tropical timber products are required to furnish the necessary information.

*Sawmill timber grading, Malaysia: need to assess risk of illegally logged timber entering the supply chain*

Certification and legality assurance systems require timber product exporting countries to 'step-up' and put in place governance, administrative and information procedures that may be beyond those that presently exist. Speaking at the meeting of the Global Timber Forum in Rome in May 2013, then executive director of the International Tropical Timber Organization[178] Emmanuel Ze Meka invited forum participants to be appreciative of the challenges certification presented for tropical timber exporters. He said that in addition to requirements like customs declarations, export permits and duties, tropical timber exporters have over the years increasingly had to contend with a wide and changing variety of information demands based on technical, environmental and phytosanitary standards, and now certification and verification of legality.

Mr Ze Meka said he was aware that these requirements arise from concerns about poor forest governance, deforestation and degradation of tropical jungles. He commented:

> In order not to be accused of being associated with these persistent problems, major consumers are increasingly using their leverage to ensure that their continued consumption of tropical timber products will not exacerbate the problems of illegal logging in the tropics as well as the continuing degradation and destruction of tropical forests.

However, Mr Ze Meka said the reality was that many of the developing tropical countries under the spotlight simply do not have adequate resources, means or capacity to fully conform to emerging certification requirements:

> The stark implication of this for producers is that access of tropical timber products to the major consuming markets of Europe, USA, Australia and elsewhere will be denied if the legal requirements of the market are not met.

He noted that while it may be possible to divert tropical timber to other markets that have not yet imposed similar requirements against illegal logging or to domestic markets in the short term, this opportunity may not have a long life. This is likely to be the case as more countries, including those that re-export to countries with legality assurance legislation begin to develop their own laws against illegal timber products or to substitute

---

178 The International Tropical Timber Organization (ITTO) is an intergovernmental organisation promoting the conservation and sustainable management, use and trade of tropical forest resources. Its members represent about 80 per cent of the world's tropical forests and 90 per cent of the global tropical timber use. ITTO was established under the auspices of the United Nations in 1986 amidst increasing concern for the fate of tropical forests.

imports of tropical timber with legal timber from other sources.

Mr Ze Meka said that in order to meet consumer country requirements tropical timber producing countries will have to devote attention, time and resources to reviewing and reforming forest governance:

> This should form the basis for the development of national legality assurance systems incorporating control of the supply chain, verification of compliance, licensing and independent auditing that meets the requirements of international markets.

The gap in capacity to implement systems for achieving legality in tropical timber producing countries still has to be addressed in some Southeast Asian countries. In some instances the costs of strengthening governance, and of establishing, maintaining and improving legality assurance systems is substantial. All this when the costs of producing and exporting legal timber products is not yet adequately reflected in the selling prices of such products. This constitutes a disincentive for producers to make further progress in achieving and demonstrating the legality of their timber products. The only certainty for tropical timber producers is that continuing access of products to major consumer markets will require a demonstration of legality. This is the justification for having to meet the additional costs of legality assurance schemes.

Thanks for sticking with me through this chapter – told you it would need to get down into some of the detail. As I said at the outset, tracking timber along the supply chain so that legality or sustainability can be asserted is technically and administratively demanding. I have attempted to illustrate that in a Southeast Asian context issues remain to be resolved. However, there is a widening recognition of the importance and value of legality and certification assurance in world markets, and a willingness on the part of Southeast Asian timber exporters to confront these challenges – we look forward to future progress.

# 16

## Is that hot enough for you?

### Climate change – coming ready or not

I reckon you don't have to be a rocket scientist – or a climate change scientist for that matter – to work out that all the pollutants from the millions and millions of tonnes of fossil fuel we have burnt since the Industrial Revolution[179] have gone somewhere up there into the atmosphere and cannot be doing us and the planet much good. Not hard to figure out.

While debate still rages over by how much average temperatures will rise as greenhouse gases build up, trapping outgoing heat there is no question that atmospheric carbon dioxide levels are rising and that this trend has implications for the climate, including the potential to seriously disrupt plant communities.

But let us step back for a minute, greenhouse gases are a natural part of the atmosphere. They include water vapour, carbon dioxide, methane, nitrous oxide and some artificial chemicals like chlorofluorocarbons.[180] Greenhouse gases absorb and reflect the sun's warmth and regulate the Earth's surface temperature.

The concentration of water vapour, the most abundant of the greenhouse gases, is highly variable. The concentrations of other greenhouse gases are influenced by human activities, particularly burning fossil fuels (coal, oil and natural gas) and also by land clearing. Once released into the atmosphere many of these greenhouse gases remain there for a long time. It is the increasing concentrations of these gases that trap heat and cause temperatures to rise.

After water vapour, carbon dioxide is the most significant of the greenhouses gases followed by methane. The simpler molecular structure of

---

179 The Industrial Revolution was considered to be the time of transition to mechanized manufacturing that occurred in the period from about 1760 to about 1840. This period included going from hand production methods to machines, new chemical manufacturing and iron production processes, the increasing use of steam power and the development of mechanised tools. The transition also included the change from wood to coal and the start of oil based fuels. The Industrial Revolution began in Britain and within a few decades spread to Western Europe and the USA.
180 Chlorofluorocarbons (CFCs) are organic compounds that contain carbon, chlorine, hydrogen and fluorine. Many CFCs have been widely used as refrigerants (in refrigerators and freezers), as propellants in aerosol applications, and as solvents. The manufacture of such compounds has been phased out because they contribute to depletion of the ozone layer in the upper atmosphere.

other greenhouses gases, like nitrogen and oxygen is not complex enough to vibrate in the infrared light range and to trap heat. This is why increases in the percentage of atmospheric carbon dioxide or elevated levels of methane can have repercussions for global temperatures.

To illustrate the point, think of the planet's atmosphere like a quilt – the more feathers in a quilt the more heat is trapped. This analogy applies to increasing amounts of greenhouse gases – additional layers of insulation that keep too much of the sun's heat trapped within the atmosphere elevating temperatures.

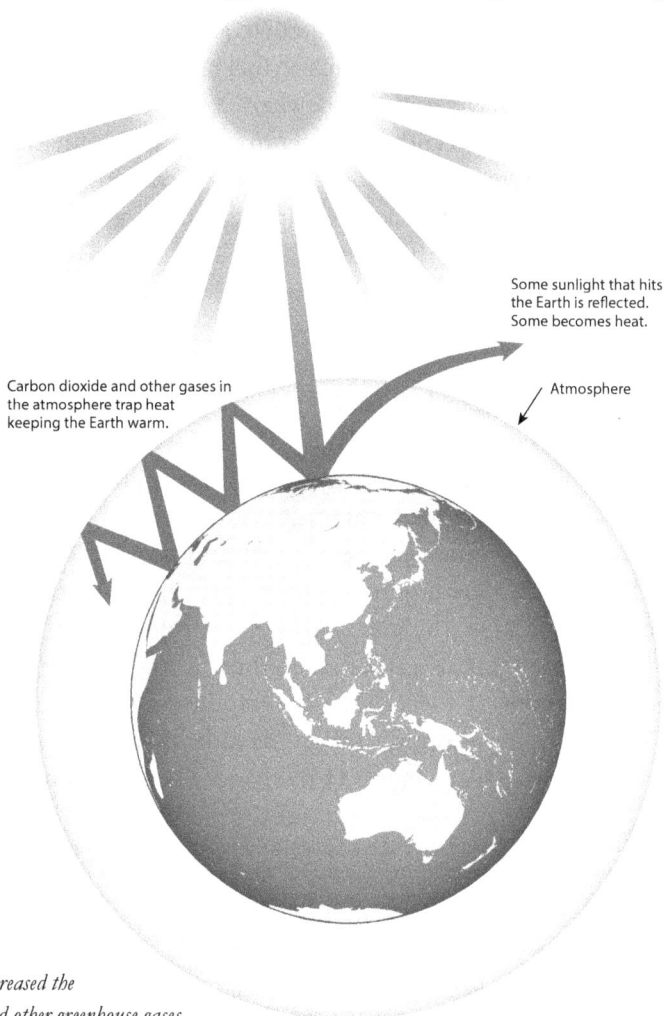

Some sunlight that hits
the Earth is reflected.
Some becomes heat.

Carbon dioxide and other gases in
the atmosphere trap heat
keeping the Earth warm.

Atmosphere

*The Greenhouse Effect:*
*Human activities have increased the*
*level of carbon dioxide and other greenhouse gases,*
*trapping more heat in the atmosphere and causing global temperatures to rise*

The magnitude of the Greenhouse Effect can be estimated using climate models. Numerous simulations have led to a consensus by the International Panel on Climate Change (IPCC)[181] that average global temperature could rise somewhere between one and three and a half degrees Celsius by the year 2100. Such a climate shift is certainly not unprecedented in the Earth's history, but the predicted pace of change – giving ecosystems and human societies only decades rather than centuries, or millennia to adapt – is certainly unprecedented.

Concern about the impacts on climate of increased atmospheric concentrations of carbon dioxide and other greenhouse gases has focused policy attention on the dynamics of carbon stored in vegetation and soils. As discussed in earlier chapters, not only does climate dictate whether forests, grasslands, or deserts exist, but a shift in the vegetation type can in turn alter local climates, and change the conditions that determine the type and pattern of vegetation present.

The ability of trees to absorb and store carbon as wood means the production of timber products makes a contribution to abating the adverse impacts of climate change. We know that these miracles of nature we call trees do more than just produce oxygen and 'lock up' carbon dioxide – the world would be a much hotter place without them.

Trees have been called the 'low tech' solution to energy conservation. Shade from trees reduces the need for air conditioning in summer, and in winter trees break the force of chilly winds, lowering heating costs. Studies have shown that parts of cities without the cooling shade from trees can become 'heat islands,' with temperatures as much as three degrees Celsius higher than surrounding areas. By using trees in cities, we are able to moderate the heat island effect caused by pavement and buildings.[182]

Yes, trees cool the air by shading and through water evaporation. They act like pumps to cycle water up from the soil back into the air. The 200,000 leaves on a large eucalyptus tree can take 40,000 litres of water from the soil and breathe it back into the air in a growing season. The cooling effect of all this water from just a single tree going back into the atmosphere is said to be

---

181 The IPCC reviews and assesses the most recent scientific, technical and socio-economic information produced worldwide relevant to the understanding of climate change. It does not conduct any research nor does it monitor climate related data or parameters.
182 For additional information see Cool Communities at: www.coolcommunities.org/urban_shade_trees.htm

equivalent to air conditioning ten rooms for 20 hours a day.[183] So common sense would suggest that solutions to the threat of climate change must include programs to plant trees, establish new forests, regenerate existing ones and keep them healthy.

*Young teak plantation: ability of trees to absorb and store carbon contributes to abating the adverse impacts of climate change*

The carbon that trees and other plants 'pull' from the atmosphere can be returned rapidly as microbes or via animals that consume leaves and fruit. In addition to breathing carbon dioxide directly back into the air as a waste product, other plant materials 'burn' the carbon compounds to fuel their own life processes. Carbon may also be stored – or sequestered – for decades or even centuries in the wood of long lived trees, in humus rich soils or peat deposits, or in manufactured timber products.

A 'complication' to the tree-carbon storage scenario arises when trees

183 For additional information see Canopy at:www.canopy.org/pages/about-trees/the-benefits-of-trees.php and Cool Communities at: www.coolcommunities.org/urban_shade_trees.htm

approach maturity and their growth slows and eventually stops – trees do not grow forever. A forest of mature trees 'locks up' large amounts of carbon, but little new growth takes place in a mature forest. To store ever increasing amounts of carbon in wood requires log harvesting and tree regeneration or replanting. Even then the wood harvested must be stored long term, or turned into durable timber products like houses and furniture.

*Wooden furniture: carbon stored long term in durable timber products like houses and furniture*

Burning wood simply releases stored carbon back to the atmosphere. However, if wood is substituted for fossil fuel, the carbon released by burning can be cancelled out by regrowing plants, causing no net overall increase in atmospheric carbon dioxide levels. Short rotation tree crops may have the potential to reduce the world's fossil fuel related carbon dioxide emissions by as much as 20 per cent.[184]

Let us step back into geological time for a moment – the climate has changed before and doubtless will change again. Indeed, some trees benefit

---

184 *The Implications of Growing Short-Rotation Tree Species for Carbon Sequestration in Canada* 1999 R. Samson, P. Girouard, C. Zan, B Mehdi, R. Martin and J Henning. Resource Efficient Agricultural Production Canada, Quebec, Canada.

from increased levels of atmospheric carbon dioxide. The four degree Celsius temperature increase that northern Europe experienced at the end of the last ice age resulted in deciduous trees flourishing at more northern latitudes. Coniferous trees also moved northward. This time, however, the climate is changing much faster than ever before, and plants are likely to have great difficulty adapting to such a rapid pace of change.

So to get to grips with climate change we should re-examine the miracle of trees – their life and death – so we are able to prevent threats to their survival. If we do not understand how forests work, it is easy to conclude that for climate change and other environmental reasons, it's better to leave trees untouched when the opposite may be true. Certainly young, healthy, growing forests do a better job for the environment climate change wise than do older ones. Plus when a tree is harvested and converted into timber products the carbon remains locked up. The wood in a typical 180 square metre timber framed house stores about eight tonnes of carbon.[185] With substantial energy required to convert iron ore into steel, the same house built with a steel frame adds about three tonnes of carbon to the atmosphere.

The more trees grow, the greater is the removal of carbon from the atmosphere and stored as wood fibre in their trunks. The only limits to how many trees can be grown are land availability and suitable climate, especially rainfall.

The European Commission estimates that on average substituting one cubic metre of wood for other building materials results in almost a tonne of carbon dioxide emission savings.[186] In addition to these savings, the ongoing 'operational' costs of a house can be reduced because of timber's thermal efficiency and its contribution to building functional performance.

There is no doubt that building design, with a strong focus on timber, is a very practical way to move the global carbon balance in the right direction. The timber processing industry could further assist by ensuring that the positive contribution forestry and timber products have to make to climate change amelioration is better recognised in policy decisions.

---

185 Useful information about building sustainability, climate change and carbon storage is set out in material from the Centre for Timber Technology and Construction, Watford, UK, including *Building sustainability with timber* and the website www.wood.for.good
186 European Commission Enterprise DG Issue 11, 2003.

*Home, Canada: carbon locked up in houses and other buildings constructed from timber products*

In terms of better understanding the potential benefits of the use of timber, rather than say plastics or steel, to climate change abatement, research conducted by Forest and Wood Products Australia,[187] shows that an overwhelming 89 per cent of people love the look of wood in their homes, buildings, furniture and flooring. However, many remain unaware of the environmental benefits of using responsibly sourced wood as part of the answer to climate change.

Forest and Wood Products Australia surveys throughout 2011 and 2012 showed that although 93 per cent of people understood that trees absorb carbon, only 52 per cent understood that choosing wood impacts positively on climate change.

---

187 Forest and Wood Products Australia (FWPA) is the forestry and wood industry's service provider, investing in research and development Australia. FWPA aims to improve the competitiveness and sustainability of the Australian forest and wood products industry through innovation, and investment in effective and relevant research and development. Visit; www.fwpa.com.au

*LVL beams: building design with a focus on timber a practical way to move the global carbon balance in the right direction*

When people talk about reducing the impact of climate change or improving air quality, they often urge the use of renewable rather than fossil fuels. Over the past couple of decades, the wood processing industry has reduced its energy consumption by more than half. Some manufacturing facilities not only supply their own energy needs, but sell surplus energy to other consumers. Today almost 60 per cent of the paper industry's energy use is self generated from bark, sawdust and sawn timber waste.[188]

As oil reserves are further depleted and petrochemical product prices escalate, the hunt for alternatives to fossil fuels is taking on a new urgency. Trees that convert simple sugars into complex chemicals are starting to look more like being part of the renewable, alternative fuels solution of the future. With the reduced availability and increased cost of fossil fuels, the world is likely

---

188  Related information in Chapter 8 of *Trees that call Australia home*, 2008 John Halkett 2008, Potts Point Publishing, Sydney, Australia.

to see a change from a petrochemical to a ligno-chemical[189] industry. Such an industry would in large part be based on biomass from trees.

So then, wood offers a vital part of the means of maintaining future living standards by using fewer non-renewable resources and less fossil fuel. The world would be more sustainable and life more certain in the decades ahead if there is greater advocacy for wood use, as well for the practice of sustainable forestry.

*Landfill wood waste: wood as a renewable energy source will reduce dependence on fossil fuels*

Carbon dioxide emissions produced by human activity amounted to a total of about 25 billion tonnes a year for the past decade, or approximately four tonnes per person.[190] Climate scientists point to the necessity of halving this rate by the middle of this century if climate change trends already apparent are to be kept within tolerable limits.

Nearly half of worldwide carbon dioxide emissions can be laid at the

189  Refers to wood based and/or derived chemicals.
190  *The Carbon Cycle and Atmospheric Carbon Dioxide* 2001 I.C. Prentice et.al www.grida.no/climate/ ipcc_tar/wg1/pdf/tar-03.pdf

door of industrial countries. The inequalities here are similar to those for fuel consumption. At 12.6 tonnes per person, carbon dioxide emissions in industrial countries are five to six times higher than in the developing countries. Across developing countries the average emissions per person is 2.3 tonnes, but with a large range of variation – from less than a tonne in the poorest countries to roughly 4.5 tonnes in those countries with increasing personal income levels. The spectrum is also wide in industrial countries – from 5.5 tonnes in Malta and Sweden to 20 tonnes in the USA. That is 200 times more than in some countries of central Africa.[191]

Because of the potential of sea level rise, global warming brings real risks to human survival. Rising sea levels will impact unevenly across the planet and may push economically weak regions and people to the limits of survival – or beyond. For example, at least 100 million people live below about a metre above sea level, which according to climate experts, is the possible level of sea level rise in the foreseeable future.

The tug-of-war over carbon dioxide emissions has centred around two interrelated agreements; the 1992 United Nations Framework Convention on Climate Change and the 1997 Kyoto Protocol. The United Nations Framework Convention on Climate Change sets out a framework for scientific and political cooperation, with the Kyoto Protocol going considerably further, and established legally binding minimum emissions reduction obligations for industrial countries.

Because greenhouse gas emission know no frontiers international consensus and action is needed to tackle global emissions levels. The central idea of the United Nations Framework Convention is that there is a common duty to: *prevent dangerous human-induced interference with the climate system*, essentially the unrestricted pumping of carbon dioxide, and other greenhouse gases into the atmosphere.

Fundamental to both these agreements is a recognition that the climate change debate has thrown up a number of questions. How much of the pace of climate change can be attributed to human influences? What

---

191  Some of the detail about climate change impacts in this chapter are taken from Chapter 9 of the very good book: *The work of nature: how the diversity of life sustains us*, written by Yvonne Baskin in 1997, Island Press, Washington DC, USA. Yvonne Baskin is a widely published acclaimed science journalist and writes extensively on the consequences arising from the present accelerating losses in biodiversity.

degree of global warming is acceptable? When will 'interference' with the climate system become dangerous, and for whom? Answers to these and similar questions will be critical for determining upper emission limits, acknowledging amongst other things that the damaging effects of climate change will not hit everyone equally. For example, rising temperatures seriously threaten the economic security and the culture of Inuit people of Canada's Arctic region. Hunting lifestyles are disappearing as traditional routes across the ice become unusable. Inuit food supplies are threatened, permafrost breaks up and igloos lose their insulating properties.

Cuts to fossil fuel use are imperative, not only to protect the atmosphere, but also to protect human rights. Millions of people may well lose what should be at the very core of their basic rights – food, water, fertile soil and a place to live. Let us be clear, the reality is that limiting greenhouse gas emissions involves a choice between human rights and consumption rights. The task of keeping temperature rises below two degrees Celsius now appears to be too great – and too threatening to the interests of those who profit from current levels of consumption.

According to climate change modelling, it is already necessary to reduce carbon dioxide emissions by 45 to 60 per cent between now and the year 2050 to contain climate change impacts to less than four degrees Celsius.[192] The Kyoto Protocol, with its commitment by industrial countries to emission reductions of around five per cent between 1990 and 2012, falls well behind this target. Also the Kyoto Protocol is severely limited and inadequate because some industrial countries – including the USA, a leading source of emissions – are outside the treaty altogether, and have demonstrated a reluctance to substantially change energy consumption behaviour. The Kyoto Protocol is also weak because developing countries have so far escaped without any emission limitations. This situation cannot remain, for emissions of 'emerging' countries will exceed the capacity of the atmosphere to absorb them, even if every industrial country were to completely cease emissions.

Negotiations in relation to a revised set of commitments under the Kyoto Protocol took place in Paris in December 2015. This widely published

192 *Climate change: A three-degree warmer world by 2050?* March 2012 IRIN Humanitarian news and analysis. (a service of the UN Office for the Coordination of Humanitarian Affairs). See: http://www.irinnews.org/report/95182/climate-change-a-three-degree-warmer-world-by-2050

gathering of world leaders and thousands of supporting experts and advocates culminated in 187 countries agreeing to limit global warming to one and a half degrees Celsius. This agreement, thought unthinkable just a few months earlier, is a substantial advance on the two degrees Celsius target that around 200 countries had agreed as a limit six years ago in Copenhagen.

At the commencement of the Paris talks countries submitted pledges to cut or curb their carbon emissions. However, collectively these pledges were not sufficient to prevent global temperatures from rising beyond two degrees. In fact several analyses suggested they may still see global temperatures rise to three degrees.

The Paris meeting resulted in countries promise to attempt to reduce global emissions down from peak levels as soon as possible. More significantly, they pledged:

> ... to achieve a balance between anthropogenic emissions by sources and removals by sinks of greenhouse gases in the second half of this century.

In simple English this means getting to "net zero emissions" between 2050 and 2100. The UN's climate science panel argued that net zero emissions must happen by 2070 to avoid dangerous global warming.[193]

In relation to text of the climate pact released at the conclusion of the Paris meeting, countries were encouraged to take action to implement and support positive incentives directed at reducing emissions from deforestation and forest degradation, and supporting the role of conservation, sustainable management of forests, and enhancement of forest carbon stocks.[194] Emphasis was placed on policy approaches that supported integrated, sustainable management of forests, while at the same time recognising the importance of actions and incentives to safeguard non carbon benefits associated with such approaches to forest management and protection.

---

193 Base on media reports including those in *The New York Times*. See: http://www.nytimes.com/ interactive/2015/12/12/world/paris-climate-change-deal-explainer.html?_r=0
194 United Nations, *Tackling Climate Change*. See: http://www.un.org/sustainabledevelopment/ climate-change/

# 17

## HUMANITY BY THE MILLIONS

### POPULATION – HISTORY, DIVERSITY AND EXPANSION

I want to convey something of the flavour of the human population across Southeast Asia in this chapter – its history, numbers, trends and impact on the jungles of the region. For the purposes of this book Southeast Asia – squeezed between the Indian Ocean and the Pacific Ocean – has a population of around 590 million, give or take perhaps ten million or so. These numbers make Southeast Asia one of the most populated regions in the world.[195] The population has increased by almost 50 per cent over the past quarter of a century – certainly a bit scary and with significant implications for the natural world.

The nineteenth and twentieth centuries have seen an extraordinary multiplication of the population right across Southeast Asia, from little more than 30 million in 1800 to 80 million in 1900, and around 525 million in 2000. A rapid decline in infant mortality and longer life expectancy from the late 1940s, plus a rise in fertility in some countries have contributed to an acceleration in population growth from the 1950s. However, to put this into context, even the recent decades of rapid population increase have left overall population densities in Southeast Asia below those of countries such as Japan, Korea, Bangladesh and India.

It is fair to say that in recent decades Southeast Asia has done well economically. It now comprises some of the most developed countries in Asia, although some still languish behind others. Today Singapore and Brunei are two of the world's wealthiest countries, while Malaysia and perhaps Thailand are in the upper middle income bracket on a world scale. The Philippines, ahead of Thailand in 1980, has experienced sluggish

---

195 Much of the discussion on Southeast Asia populations is taken from *The Population of Southeast Asia* (ARI Working Paper No.196) 2013 by Gavin W. Jones, Asia Research Institute, National University of Singapore.

economic growth in recent decades, and has fallen behind Thailand. Most agree that the three poorest countries in the region remain Cambodia, Laos and Myanmar.

Although prior to the twentieth century a number of medium sized cities existed across Southeast Asia, as did some densely settled rice growing regions, much of Southeast Asia remained sparsely settled relative to parts of Asia. This low population density has influenced the character of much of the region, with relatively weak national cohesion and commerce, and with communities relying on subsistence shifting cultivation.

However, in a further example of the escalating influence and ascendancy of the East, there has, over the course of the last century, been a dramatic reversal in the demographic balance between Southeast Asia and Europe. Take Indonesia as an example; at the start of the twenty first century, the Indonesian population exceeded that of Russia – the largest European country – by almost 100 million. There are now more Vietnamese and Filipinos than Germans. Thailand, a medium sized Southeast Asian country has a larger population than either Italy or the United Kingdom.

As indicated the economic revolution across Southeast Asia has been accompanied by extraordinary rates of population growth, especially during the second half of the twentieth century. At the end of the colonial era, about 1950 – the population of Southeast Asia was only one third as large as that of Europe – the home of the colonial powers that ruled most of Southeast Asia for the first half of the century. Now the combined population of the Southeast Asian countries of Indonesia, Laos, Malaysia, Myanmar, Philippines, Cambodia, Singapore, Thailand, Vietnam and Brunei is projected to approach 600 million by the end of the second decade of the twenty first century and Southeast Asia is projected to be home of more than 750 million by the middle of the century. You get the picture.

Today major features of Southeast Asian cities are shopping malls, congested roads and pervasive smog from urban industries and transport. The subsistence low tech economies of Southeast Asia's past have been replaced by dynamic economies producing industrial and electronic goods, clothing and footwear and home and industrial appliances for world markets.

Today Southeast Asian countries boast major metropolitan cities, including Jakarta, Bangkok, Singapore, Manila, Kuala Lumpur and Ho Chi Minh City. From their humble beginnings as early trading centres, or centres of colonial administration, these places have been transformed into global cities. In recent decades the appearance and economic structure of many Southeast Asian cities has increasingly resembled that of modern cities of the West.

It may surprise you, but well beyond the reach of modern cities and rural lowlands of Southeast Asia are remote, often mountainous areas. Wide expanses of ocean and variations in geography have created niches that have facilitated an incredible diversity of cultures. Remote regions are frequently populated by ethnic tribal minorities rarely integrated into national language, social fabric or commerce networks. In addition to these ethnic minorities most Southeast Asian countries are also home to immigrant communities, the largest being of Chinese descent. Chinese migration to Southeast Asia began well before modern times, although major waves of migration from China to Southeast Asia occurred in the late nineteenth and early twentieth centuries, during the period of colonial rule.

*Ethnic Chinese fish farm: migration to Southeast Asia began long before modern times*

Right across Southeast Asia for more than a millennium there has been frequent contact, trade, migration and social exchange from other parts of Asia and elsewhere. Over the past five hundred years, European merchants, along with adventurers and missionaries arrived in the region. Cultural influences from outside Southeast Asia have blended with local religious traditions, economic activity and politics.

Turning briefly to agricultural activity, the traditional agricultural crop of Southeast Asia is rice grown in dry fields and in rain fed or irrigated fields. Since wet rice – grown in irrigated fields – is a much more productive crop, there has been a substantial trend towards wet rice cultivation. Over the last century or more, most fertile lowland areas have been settled, and irrigated rice fields are now a dominant feature of the landscape. The scale of human effort necessary to transform tropical jungle and swamp into irrigated rice fields is enormous and only possible with higher population densities and coordinated effort to construct irrigated systems and related infrastructure.

*Transplanting rice seedlings: effort necessary to transform jungle and swamp into irrigated rice fields is enormous*

Countries across Southeast Asia practice different religions. Islam is the most widely supported religion in the region numbering about 240 million followers, primarily in Indonesia, although there are wide variations in practices and beliefs. Peninsula Malaysia shares many religious, cultural and linguistic traditions with Indonesia. Buddhist beliefs, institutions and traditions have had a significant influence on the social patterns and culture in Myanmar, Thailand, Laos, Cambodia and Vietnam. Since the sixteenth century Christianity has been the dominant religious tradition in the Philippines.

In spite of these broad generalisations there is considerable religious diversity across the region. There are Muslim populations in Singapore, southern Thailand and the southern Philippines. There are small Christian enclaves throughout the region and Hinduism is the major religion in Bali, and among the Indian minority populations of Malaysia and Singapore.

For a moment let us go back to the beginning population wise to put present numbers in Southeast Asia into some sort of context. We know that the human species is an African one. Africa is where we evolved, and where we have spent the majority of our time on Earth. The earliest fossils of modern man, or *Homo sapiens,* appear in the fossil record of Ethiopia dated at around 200,000 years old.

According to fossil finds we humans only started to leave Africa between 60,000 and 70,000 years ago. What set this exodus in motion is uncertain, but it is thought it had something to do with major climatic changes happening at that time – a rapid cooling in the Earth's climate – driven by the onset of the last Ice Age. This cold snap would have made life difficult for our African ancestors and the paleontological[196] evidence points to a sharp reduction in the human population around this time. In fact, the human population is considered to have likely dropped to fewer than 10,000 – we were just hanging on by a thread.

Once the climate started to warm again, early humans came back from near extinction. The population grew and explorers ventured out beyond Africa. Early wanderers expanded along the coast to India, then reached Southeast Asia and Australia about 50,000 years ago. This started an amazing dynasty of settlement and expansion that continues to this day.

---

196 Palaeontology is the study of the forms of life existing in prehistoric or geologic times, as represented by the fossils of plants, animals and other organisms.

Europeans began to make their presence felt in Southeast Asia just a few hundred years ago. Since early times, the Indonesian archipelago had been known as a source of spices. The spices in question were cinnamon, cloves, mace and nutmeg which Europeans could not grow, but valued to enhance the taste of their otherwise bland diet. For centuries the spice route had run from the Indian Ocean up the Red Sea, or overland through Arabia and Anatolia.[197] By the middle of the fifteenth century the lucrative final leg leading into Europe was tightly controlled by the Turks and the Venetians.

Until the twelfth century, many Europeans believed that the only way to the East was overland. The Portuguese realized that if they could find an alternative route down the west coast of Africa and round the Cape of Good Hope into the Indian Ocean, they would be able to dominate the spice trade.

Portuguese mariner Bartolomeu Dias rounded the Cape of Good Hope in 1488, but was forced to turn back by his crew. Nine years later, under the patronage of Portuguese King Henry the Navigator Vasco da Gama rounded the Cape of Good Hope, reaching India in 20 May 1498. While half the men on da Gama's expedition did not survive the voyage the way now lay open for sea expeditions further east. Da Gama himself died of malaria during his third trip to India in 1524.[198] The Portuguese spread eastwards to Malacca[199] in what is now Malaysia, and in 1511 three ships sailed for the Moluccas of Indonesia, famous for their spices, and still called the Spice Islands today.

Ferdinand Magellan[200] sold the idea of a shorter voyage to the Spice Islands to the King of Spain. He set off in his ship the *Victoria* and returned laden with spices, though Magellan himself was killed in the Philippines.

The sixteenth to the nineteenth centuries saw the Dutch, French,

---

197 Also call Asia Minor, is the western most portion of Asia that today makes up the majority of Turkey bounded by the Black Sea to the north, the Mediterranean Sea to the south and the Aegean Sea to the west.
198 The discussion on early European voyages to Southeast Asia is based on: *Civilization – The West and the Rest*, 2011, Niall Ferguson Allen Lane, Published by Penguin Group, Australia.
199 The Moluccas also known as the Maluku Islands are an island group within Indonesia. Geographically they are located east of Sulawesi, west of New Guinea and north and east of Timor. Historically the islands were known as the Spice Islands by the Chinese and Europeans.
200 Ferdinand Magellan (1480-1521) served King Charles I of Span in search of a westward route to the Spice Islands. Magellan's expedition of 1519–1522 became the first expedition to sail from the Atlantic Ocean into the Pacific Ocean the passage being made via the Strait of Magellan, and the first to cross the Pacific. His expedition completed the first circumnavigation of the Earth although Magellan himself did not complete the entire voyage, but was killed in the Philippines.

Spanish and British fighting for dominance in Southeast Asia. The Dutch founded the Dutch East India Company, and consolidated their presence in the region on the Indonesian islands, while the British founded the British East India Company and controlled Malaysia. During the early colonial era colonists made a substantial contribution to the regional economy by creating plantations using plant species, such as tea, coffee and rubber, not native to the region.

*Fifteenth century Portuguese caravel: by finding a route around the Cape of Good Hope into the Indian Ocean the Portuguese were able to dominate the spice trade*

Although driven mainly by trade in spices and later plantations these European colonies also attracted scientists in search of new biological 'treasures'. One of region's British governors was Sir Thomas Stamford Raffles, a gifted administrator and naturalist who subsequently founded Singapore. The largest flower in the world, *Rafflesia* bears his name.[201]

---

201 *Rafflesia* is a genus of 28 species of parasitic flowering plants found in Southeast Asia. In some species, such as *Rafflesia arnoldii* the flower may be over 100 centimetres in diameter and weigh up to ten kilograms making it the largest flower in the world.

So what have we learnt in this chapter that is relevant to the central theme of this book? Certainly the human population across Southeast Asia has grown like topsy and shows no signs of slowing. More people and improved living standards will place even more pressure on the land. Jungles will continue to be cleared for settlement and agricultural development. Unfortunately these and other economic 'pressures' will drive additional adverse impacts on nature conservation – habitat loss, declining biodiversity and species extinction.

All together the impacts of growing humanity are bad news for jungle survival. Population expansion will need to be managed, requiring government commitment and political courage. We will have to wait and see.

# 18

## DOING LESS TO DO MORE

### TIMBER USE – INNOVATION AND ADDING VALUE

I want to talk about more astute ways to utilise timber sourced from tropical jungles in this chapter. The more frugal, value added use of this wonderful resource holds part of the secret – part of the solution – to saving jungle trees. Yes, in what perhaps seems to be a contradiction: using wood may save trees.

We will get to the detail later, but the essential point in opening this discussion is to emphasise the merits of utilising tropical timber for applications that recognise or need its physical, decorative and engineering attributes. Also part of the saving-the-jungle-trees solution is not to use tropical timber in situations where plantation based, solid or engineered wood products[202] will do the job just as well.

Sorry to return to Australia again, but for a moment let us draw a parallel with internationally renowned Australian hardwood timbers like jarrah, karri, red gum, blackbutt[203] – and others – the list is a long one. Today such timbers are used predominantly for decorative purposes, like flooring, panelling and furniture, whereas in days past massive Australian hardwood beams were used as major structural components in commercial construction. Great

---

202 Engineered wood products are made of wood based components remanufactured into structural building products certified and complying with the requirements, structural engineering codes and standards. Engineered wood products are usually seasoned, dimensionally accurate and able to be produced in long lengths and large end sections to replace solid timber in critical structural applications.
    Engineered wood products include Laminated Veneer Lumber (LVL); I-beams (I-shaped engineered wood based structural members made up of a top and bottom flange of solid timber or LVL separated by a vertical web of structural plywood or other wood-based product); Glued Laminated Timber, often referred to as Glulam (an engineered structural product produced by bonding together a number of graded, seasoned and mostly finger jointed laminates with a proven structural adhesive to form a solid member), and plywood.
203 Jarrah (*Eucalyptus marginate*), karri (*Eucalyptus diversicolor*), red gum, (*Eucalyptus camaldulensis*) and blackbutt (*Eucalyptus pilularis*). As an example, jarrah was renowned as a versatile all-purpose timber. It has also been widely used as a utility timber because of its strength, durability, availability and resistance to fire. Before the development of bitumen and concrete road surfaces, many famous thoroughfares around the world were paved with jarrah blocks.

jarrah and karri timber sections formed the skeletons of Australia's (and other countries) industrial and commercial buildings for well over a century. Timber was used for structural work because it is availability and low cost. In such applications the special attributes of timber from natural forests was overlooked and undervalued. Today engineered beams, plywood bracing, particle board and prefabricated panels are used for structural work instead. Indeed, the great hardwood beams of yester year are being recovered and recycled into trendy flooring, furniture and for architectural interior designs.

Also today there is a broadening acknowledgement that finely crafted native timber furniture, renowned for its quality, beauty and durability, is an asset in any home and that buying such furniture is not only a guarantee of a fine product, but a good choice for the environment.

Leading United Kingdom architect Alex de Rijke[204] argues that if the nineteenth century was the century of steel, and the twentieth century the century of concrete, then the twenty first century will be about engineered timber. Australian construction industry advisor David Chandler adds that while timber is a foundation construction material it has not generally been considered as a viable alternative to concrete, steel and masonry. "This may be about to change. Engineered wood products could be at the cutting edge of that change," he said.[205]

Some architects say that the modern timber industry must rethink how it can better process timber into higher value products, particularly engineered products for prescribed applications. There remains criticism by some that the industry has failed to adequately move with the times and continues to promote traditional products that lack innovation and application to modern design, life style trends and construction needs.

Prominent Australian architect Chris Howe is an advocated of timber use, but a critic of industry's efforts. He said:[206]

204 Alex de Rijke is a founding director of the United Kingdom architectural practice dRMM. His work is well-known for innovative construction technologies and materials. He has taught at the Architectural Association, London; the Aalto University, Helsinki and as Guest Professor at the School of Architecture in Düsseldorf. The Royal College of Art has appointed Alex de Rijke as Dean of the School of Architecture.
205 *Wood and Australia's 2023 construction road map*, 2013 David Chandler. See: www.constructionedge.com.au
206 As outlined in Chapter 9 of *Trees that call Australia home*, 2008 John Halkett 2008, Potts Point Publishing, Sydney, Australia.

Except in some areas, such as engineered structural components and laminates, the timber industry is losing market share to aluminium, masonry and steel. Building design is now a more competitive and complex market so the timber industry must provide integrated design solutions and strategic partnerships.

Howe urges the timber industry needs to develop a promotional and marketing campaign on the environmental and economic benefits of timber products:

> Green design is no longer an alternative, but a prerequisite for all projects. The objective is for architects and other design professionals to promote and specify wood as a quality, sustainable, clean, friendly and living product.
>
> Sustainability is now considered smart living, an everyday consideration in the construction market. Sustainability therefore means smart business.

A compelling lesson learnt with direct application to timber sourced from the jungles of Southeast Asia revolves around taking advantage of innovation in advanced engineered wood products. Better utilisation should be about appropriate applications and not using remarkable tropical timbers when more readily available plantation based alternatives will suffice for the job at hand.

The current drive for energy efficiency and ecologically sensitive design will become a significant driver of innovation in building materials and construction methodologies in the future. It is vital therefore to stress that timber construction is substantially less environmentally harmful than the use of aluminium, steel and cement based products.

The challenge facing the timber industry is to provide solutions that instil confidence in building designers, together with an improved understanding of timber specifications and encouragement of the use these products. Overall, timber product manufacturers need to move in the direction of higher levels of technical innovation and to embrace the development of new engineered products across a wide range of applications.

*Cross laminated timber panels: energy efficiency and ecologically sensitive design will be 'drivers' of innovation in building materials and construction methodologies*

We have already seen that trees have breathed life into humanity. Forests and jungles have been a source of food and fuel for our ancestors and have provided the material for building primitive shelters and stockades. They were the stuff from which weapons were fashioned for hunting and defence. A tree is the most durable, constantly growing and self repairing store of resources we know.

Forests and the timber they provided allowed ancient civilisations to journey onto the sea to discover new lands. The earliest floating craft were simple, wooden dugouts, bark canoes or rafts made from logs lashed together. These boats – if you can call them that – permitted our forebears to fish and to travel. Remarkably some small fragile rafts and canoes undertook amazingly long sea voyages.

Among very early seafaring civilisations the Phoenicians[207] were particularly successful. From their homelands in ancient Canaan (now Lebanon), they travelled across the Mediterranean establishing trading posts throughout the region.

While other materials can provide shelter, warmth and comfort, none are renewable like trees. Other resources frequently require considerable energy to transform them into usable products. For instance, the manufacture of clay into bricks; petrochemicals into plastics; iron ore into steel, and bauxite into aluminium all need the application of substantial amounts of energy.

On the other hand wood is a remarkable creation. It can be sawn and carved, nailed, glued, pressed, and heated. It can also be chemically broken down into cellulose, lignin and other compounds. In addition to timber products wood can be made into paper, foods, lubricants, medicines and textiles.

Throughout history we have always used trees for energy. Whether as firewood to keep us warm and cook, or today as fossil fuels, peat, coal, natural gas and oil – formed from the rich biomass of forests and jungles – slowly compressed beneath layers of earth over millions of years. When these fossil fuels are depleted, trees will continue to be a critical source of energy for future generations.

As trees grow they store more and more energy. As an example from that neat little book: *The Miracle of Trees* by Olavi Huikari that I have mentioned earlier, a cubic metre of tree trunk can in energy terms equal 220 litres of oil. Indeed, one hectare of mature forest can contain as much energy as 60,000 litres of oil. Among other uses, tree biomass can be used to generate electricity. So a forest is like a charged battery that can be worth much more per hectare in energy and other 'new generation' products than just as timber alone.

---

207 The Phoenicians were the direct descendents of the Canaanites of the south Syrian and Lebanese coast who, at the end of the second millennium BC, became isolated by population and political changes. The name derives from the Greek, *Phoinikes*, referring to the purple coloured dye which the Phoenicians extracted from the murex shell and with which they produced highly prized textiles.

The Phoenicians were fiercely independent maritime traders and by the late eighth century BC, had founded trading posts and colonies around the Mediterranean, the greatest of which was Carthage on the north coast of Africa (present day Tunisia).

*Extractive chemical research: a forest is like a charged battery that can be worth more in energy and other 'new generation' products than just as timber alone*

Returning to more traditional timber use, over 90 per cent of all timber used in Europe comes from European forests[208] that are increasing in size. Between 1990 and 2005, the forested area of Europe grew by 13 million hectares. That is an area roughly equivalent to the size of Greece. While deforestation, as we have seen, is a serious problem in tropical regions, the forests in some temperate areas are expanding. Good forest management is part of this happy circumstance in Europe, as well as the planting of trees.[209]

208 IIED, Using Wood Products to Mitigate Climate Change, 2004
209 *Building sustainably with timber* Centre for Timber Technology and Construction, 2004 London, UK.

A measure of the amount of energy used in product manufacturing processes is called embodied energy. Embodied energy is a useful way to compare the environmental 'friendliness' of various building materials. Compared to other common building materials, such as steel, aluminium and concrete, timber not only stores carbon, but it uses much less embodied energy in its manufacturing than other building products. In simple terms, a concrete slab floor uses 60 per cent more embodied energy than a timber floor; double brick walls use almost five times more than weatherboards and an aluminium window uses half as much energy again than an equivalent timber window. The substitution of timber elements for more energy consuming products in the building process results in substantial energy savings and corresponding reductions in greenhouse gas emissions.

Buildings are seldom constructed of a single material, so the embodied energy of a structure depends on the mix of materials used. Research continues to refine the detail, but the construction of a brick veneer house with timber framing generates savings of about nine tonnes of carbon dioxide per house compared to double brick construction alone.[210]

A vital aspect for producer countries is to improve the economics of tropical jungle retention by 'moving up the value chain' – or in other words – produce and sell more sophisticated valuable products than just logs or sawn timber. The maths is simple enough, the more value that is added to the product, the less sensitive is the price paid for logs. So you can pay a lot more if you make say a table and chairs rather than sell sawn timber. Also of course, you can employ people, generate associated business activity and support local economies.

To illustrate this important point further let us look at Myanmar. Although some finger jointed panels, furniture, parquet flooring, marine decking, doors and door frames are produced and exported, as are curved chairs and garden furniture, Myanmar essentially operates towards the bottom of the value chain. Data suggests that the country produces 1.4 million tonnes of teak and hardwood tropical timber annually, exports about 800,000 tonnes and earns $400 million. By contrast, Malaysia produces 300,000 tonnes of timber and earns $6 billion for wood product exports. This means that

---

210 *The environmental credentials of production, manufacture and re-use of wood fibre in Australia* 2001 Attiwill P., England J. and Whittaker K. A report prepared for the Department of Agriculture, Fisheries and Forestry, Australia.

Myanmar generates only $500 per tonne while Malaysia benefits by $20,000 for every tonne exported. This staggering difference is explained by the fact that much of Myanmar's wood based businesses are essentially 'low-tech', like sawmills and log exports while Malaysia has climbed up the value added chain exporting more wooden furniture that basic timber products.[211]

Until a ban on log exports was implemented from April 2014 logs, especially teak logs have been a substantial export from Myanmar even though the Myanmar National Development Plan[212] has targeted regular reductions of log exports and increasing supply to domestic wood based industries. The Myanmar tax system also tends to discourage export of finished products as there is a ten per cent value tax on the export of such items.

*Forté apartment building, Melbourne, Australia: at 32.2 metres it was the tallest timber apartment building in the world at that time*

211 Data from: *Myanmar forestry outlook study*, 2009 Asia-Pacific forestry sector outlook study 2, Working Paper No. APFSOS II/WP/2009/07, Khin Htun Food and Agriculture Organization, Bangkok.
212 The Myanmar Government National Development Plan covers the country's economic and social development over the next 30 years. The plan includes strategies for poverty alleviation, human resource development, investment and trade sector development, industrial, financial sector and currency development. The development strategies in the plan focus on five key areas during the development period being tourism, electricity and power generation, communication and transport, mining and natural resources, as well as property development. In the period up to 2017, the plan targets an annual average GDP growth of 7.7 per cent (from Investvine: www.investvine.com).

To provide a bit more context let's return to the discussion about the emergence of engineered wood products. These products are evolving rapidly to offer characteristics that match those of steel and concrete. The pace of these changes is most evident in the northern hemisphere where a supply chain exists that incorporates sophisticated fixings, glues, fire retardants, composites, and the use of the latest computer based design and production systems.

Countries like Switzerland, Norway, Germany, Austria and Finland are leaders in innovative, industrialised engineered wood products construction systems. Related markets have established supply chain linkages between timber product manufacturers, architects and builders able to offer a wide range of engineered and value added house and commercial construction options.

I will attempt to illustrate this point by using Cross Laminated Timber or CLT as an example. Sustainability and a reduced environmental 'footprint' have been key drivers behind the widening use of CLT. CLT manufacturing is fundamentally a process whereby timber is aligned at right angles in alternative layers then glued and pressed. This process produces CLT panels as a viable alternative building material to concrete and steel.

Apologies again for using an Australian example, but here we go. In mid 2013 the *Forté* building at 32.2 metres was constructed using CLT panels. It was the tallest timber apartment building in the world at that time. Built by Lend Lease[213] in their Victoria Harbour precinct in Melbourne and aspiring to be the first *5 Star Green-Star As Built*[214] certified residential building in Australia.

CLT provides similar levels of structural performance to traditional concrete buildings while delivering better thermal performance, and in the case of the *Forté* building reducing the building's projected life cycle carbon dioxide emissions by around 1400 tonnes.

---

213 Lend Lease is a major Australian company developing and funding significant building and infrastructure projects, including shopping malls, residential high rise residential communities, commercial developments, hotels, hospitals, roads and bridges.
214 For information see the Green Building Council of Australia (GBCA) website: https:// www.gbca.org.au/ The GBCA was established in 2002 to develop a sustainable property industry in Australia and drive the adoption of green building practices. This includes the Green Star environmental rating system for buildings. Green Star rating tools are intended to assist the developers and the construction industry reduce the environmental impact of buildings, improve occupant health and productivity and achieve real cost savings, while showcasing innovation in sustainable building practices.

In addition to the environmental benefits, the *Forté* building combines other sustainable initiatives, such as LED[215] lighting and 'smart' electricity consumption. Using rainwater tanks, *Forté* collects rainwater from the roof and uses it to supplement toilet flushing and supply the building's fire suppression system. Bathrooms are fitted with water efficient showerheads and taps, and kitchens with water and energy efficient dishwashers and washing machines. The building also boasts a car sharing system.

Lend Lease's chief executive officer Mark Menhinnitt said the *Forté* project:

> Unlocks a new era for sustainable development by offering a viable alternative to traditional construction options that are carbon intensive.

He said CLT was the most significant form of innovation in construction technology that Australia had seen in many years.

> CLT will transform the construction industry by introducing a more efficient and environmentally-friendly construction process that has never been undertaken before. Lend Lease is aiming to develop 30 to 50 per cent of its apartment pipeline using CLT.
>
> With an increasing number of people moving to urban areas, this innovation is timely given the urgency to create liveable, sustainable cities that are climate positive. The adoption of green technologies, materials and construction processes, like CLT, means we are closer to achieving this.

Being largely based on pre-fabrication, the CLT construction process involves less material on site and was cleaner, simpler and faster. Commercially, a critical factor is construction time and CLT pre fabrication means all the main penetrations and cut outs were already machined, plus fixing into timber is a lot easier than fixing into concrete, so for electricians, plumbers, plasters and other trades, it's a much easier job than working with concrete.

Because of its acceptance by architects and builders CLT is now being widely incorporated into high rise development design and residential housing throughout the United Kingdom, Europe and Australia. Globally it is the fastest growing timber product, with a current annual increase in demand of 30 per cent. It is anticipated that this trend will be maintained as confidence in design, engineering and planning approvals related to CLT use increases.

---

215 LED stands for Light Emitting Diode, a semiconductor device that converts electricity into light.LED lights are the latest technology in energy efficient lighting, using approximately 85 per cent less energy than other types of lighting.

Timber has the capacity to 'tick the boxes' in lowering on site construction labour hours by 30 per cent and the capacity to drive on site construction time down by half.[216] So now that the first batch of high rise buildings have been constructed and more are on the drawing board CLT use in commercial and high rise residential applications is predicted to accelerate.

Turning to the other important issue for tropical timber use – that of frugal, value added applications. This is illustrated by the example of a manager's house and guest accommodation built, as it happens, for a forestry company managing 100,000 hectares of Borneo forest. This zero energy building designed by Malaysian born architect Ken Yeh provides accommodation for the manager and guests of the company. A collaboration between architecture, forestry, botany and anthropology, the design is based on a modular system, using local timber and small timber sections to overcome the difficulties of a remote location, lack of infrastructure and few skilled builders. Yeh said:

> The design has been influenced by the vernacular longhouses of the area with their frugal timber construction and by Henry Thoreau's book: Life in the Woods[217] with its message of simple living and self sufficiency.

The house is low cost, autonomous, with solar electricity, bio gas units and rainwater collection. Despite the tropical latitude the indoor temperature of the house peaks at 26 degrees Celsius at midday, a full eight to ten degrees lower than the outdoor environment. Yeh added:

> It is also a symbol of craft, care and environmental stewardship; an example of what can transpire when design thinking meets difficult circumstances. Sustainability is an issue of pressing urgency, yet it can also be a buzz word with little substance. Put simply, we believe it is the overarching concept of using less to do more.
>
> "By understanding the jungle, I can create environments that are sensitive and fit correctly into the landscape. We do a lot of upfront work and this ethos has served us well in our projects.

---

216 *Wood and Australia's 2023 construction road map*, 2013 David Chandler. See: www.constructionedge.com.au

217 Life in the Woods is an American book written by Henry Thoreau in 1854. The book is part personal declaration of independence, social experiment, voyage of spiritual discovery and manual for self-reliance. It details Thoreau's experiences over the course of two years at Walden Pond in woodland in Massachusetts living a simple life without support of any kind. By immersing himself in nature, Thoreau hoped to gain a more objective understanding of society through personal introspection.

Yeh defers to the American environmental activist Wendell Berry[218] who said:

> We could say, then, that good forestry begins with the respectful husbanding of the forest that we call stewardship and ends with well made tables and chairs and houses; just as good agriculture begins with stewardship of the fields and ends with good meals.

Singapore based managing director of design and contracting company Venturer Pte Ltd, Kevin Hill is 'on-the-money' with his actions and comments, both about using pine instead of tropical timber and the utilisation of tropical hardwood for its special design and performance attributes.

He says of pine:

> We use treated softwood, such as southern yellow pine from the US, or radiata pine from New Zealand for framing, joists and rafters – internally and externally. It is wasteful to use increasingly rare, visual grade hardwoods for concealed framing. Unless you are employing their specific structural properties, it does not seem warranted.

In relation to engineered wood products, Hill is a leading developer and advocate for their use in Southeast Asia for demanding structural and outdoor projects. He has developed a merbau based glulam[219] product that has wide application and appeal.

Because merbau is considered fire resistant it has the potential to outperform even steel in a fire, especially with the application on a fire retardant coating. Hill added:

> Merbau glulam timber outperforms European and American glulams because often when you are creating a curved, elliptical or free form structure, maintaining light sections makes the effect even more striking.

Hill has created glulam laminates that are as thin as four millimetres with a radius as small as 350 millimetres – outperforming standard rules-of-thumb for cold bending by more than 200 per cent.[220]

---

218 Born in 1934 Wendell Berry is an America activist, academic, critic, and farmer. He is a prolific thinker and author of novels, short stories, poems and essays.
219 Glulam or glued laminated timber is a type of structural timber product comprising a number of layers of timber bonded together with durable, moisture-resistant structural adhesives. By laminating a number of smaller pieces of timber a single large, strong, structural timber product is manufactured. These structural products can be used in a number of aplications, including vertical columns or horizontal beams as well as curved, arched or free form shapes.
220 For further information see: *Warming a cold palette – more timber elements appearing in SEA design.* In *Timber and Design International*, First quarter 2013, Pages 22-24

# 19

## ONLY PRACTICAL SOLUTIONS NEED APPLY

### SUSTAINABILITY, CONSENSUS, ECONOMICS, INNOVATION AND SCIENCE

Now, in this chapter, we need to get to practical solutions aimed at sustaining the jungles of Southeast Asia. Obviously such solutions must be based on the reality of current populations and economic trends in this modern, diverse and progressive region. Even though the evidence that jungles are in serious trouble is irrefutable, we need to be realistic, we need to be pragmatic, and we need to be optimistic.

We need to acknowledge that without deliberate action the jungles of Southeast Asia will continue to erode and deteriorate. They may soon reach a 'tipping point' that triggers a rapid slide into extinction of several iconic animals; a traumatic loss of biodiversity, and an inability to continue to support dependent rural and tribal communities. The long term consequences of such an outcome will not only be environmentally devastating, but an economic and social disaster. The survival of the jungles – a special part of the heritage of the planet and of the cultural and economic values they contribute – is at stake. Yes, we all have a responsibility to work for their survival and future.

So as discussed what we are looking for is a practical set of solutions based around realistic measures that will enhance the prospect of sustaining the jungles. This set of solutions, together with an appreciation of their natural, heritage, national identity and commercial values, may mean that Southeast Asian jungles, together with their cargo of plants, animals and human communities will endure long after we have departed.

In earlier chapters we tracked over the issues driving jungle destruction – large scale land conversion; unsustainable and insensitive logging, fragmentation and loss of biodiversity – I am not going to regurgitate these issues again suffice to say that solutions lie in a number of actions. Sensitive timber harvesting, with an assurance of legal log supplies is a vital step, as is continuing to advance initiatives directed at sustainable forest management.

The 'case' for the jungles also lies in more frugal, value added utilisation of tropical timber, including progress on the development of engineered wood products, furniture, and other higher value products and applications. This is about encouraging timber product innovation and wood processing investment, and also noting the commercial opportunities associated with carbon storage and green star ratings schemes.[221] It is also about extending and improving the management of tree plantations to take some of the timber demand 'pressure' off jungles.

*Harvesting balsa logs, East New Britain, PNG: a case of more frugal value added utilisation of tropical timber*

In terms of moving wood processing 'up-the-value-chain', encouraging signs are apparent in some Southeast Asian countries, where a pleasing trend is a shift from being exporters of logs and less processed timber, to being exporters of more value added items, like wood based panels, plywood,

221 Green star rating schemes are broadly comprehensive, national, voluntary environmental rating systems that evaluate the environmental design and construction of houses and other buildings. Green star rating schemes have been developed for the residential and commercial building and property industries in order to establish a common set of measurements for 'built' environment sustainability; to promote integrated, holistic design, and to raise awareness of the benefits of sustainable design, construction and urban planning.

engineered wood products, components and furniture. The 2014 log export ban in Myanmar and the acceleration of the planned log export ban in PNG underscore this welcome trend.

This shift to an emphasis to domestic value added production is reflected in a review of International Tropical Timber Organization data that shows the total volume of world exports of basic tropical timber products – comprising logs and sawn timber – has fallen from 70 million cubic metres in 1995 to 52 million cubic metres in 2011. This represents a decline in the share of world exports from 22 per cent to just 13 per cent over the same period.[222]

As we have seen agricultural, industrial and urban encroachment continues apace and presents a serious challenge to jungles across Southeast Asia. Unfortunately it remains a reality that there are still too few examples of an even handed approach to jungle retention where jungle conservation is integrated into broader land use planning and where tradeoffs are prescribed between jungle attributes and management, and a range of other economic goals.[223]

Also as yet pursuing economic outcomes that recognise value and allow trade in environmental services,[224] including carbon storage and trading are not well understood and developed. However, the provision of what are generally described as 'ecosystem services' is gaining prominence so we might reasonably expect to see increasingly large tracts of tropical jungle and other natural forest types being managed for these services in the future.

The issue of the relationship between economic development and environmental degradation was placed prominently on the international agenda with the convening of the United Nations Conference on the Human Environment held in Stockholm in Sweden in 1972. By 1983, mounting concerns about global environmental 'stress' resulted in the Brundtland Commission[225] making the term 'sustainable development' part of modern

---

222 *What is happening to the tropical timber trade?* 2013 Amha Bin Buang, International Tropical Timber Organization. Address to the 3rd High Level Market Dialogue: The new era of Indonesian legal timber products to meet global markets, Jakarta, Indonesia.
223 *Asia-Pacific Forests and Forestry to 2012* Report of the second Asia-Pacific forestry sector outlook study 2010 Food and Agriculture Organization of the United Nations, Bangkok, Thailand.
224 Conservation of biological diversity, maintenance and improvement of watershed values, combating desertification and land degradation, and climate change mitigation and adaptation are key ecosystem services provided by jungles.
225 Formally known as the World Commission on Environment and Development (WCED), the Brundtland Commission's mission was to attempt to unite countries to pursue sustainable development objectives.

vocabulary. Essentially this marriage between previously strange bedfellows – economic and environmental outcomes – is now central element of acceptable natural resource management. Plus, at a time when deforestation and degradation are resulting in significant environmental damage, the role of forests in climate change mitigation is becoming centre stage in global climate change politics and discussion.

To some degree the opportunity for forestry to have a positive role in climate change abatement and the provision of environmental services depends on making progress in arresting deforestation and forest degradation. The Reduced Emissions from Deforestation and Degradation (REDD) scheme first mentioned in Chapter 13 envisages payments to forest owners and other parties for trading carbon stored in trees and for conserving forests. However, while REDD schemes seem to offer enticing prospects[226] there is still uncertainty as to how such schemes will be developed and implemented, and to what extent they will become an important component of climate change mitigation and forest protection. Stepping up tree planting to enhance carbon storage is another valuable strategy.

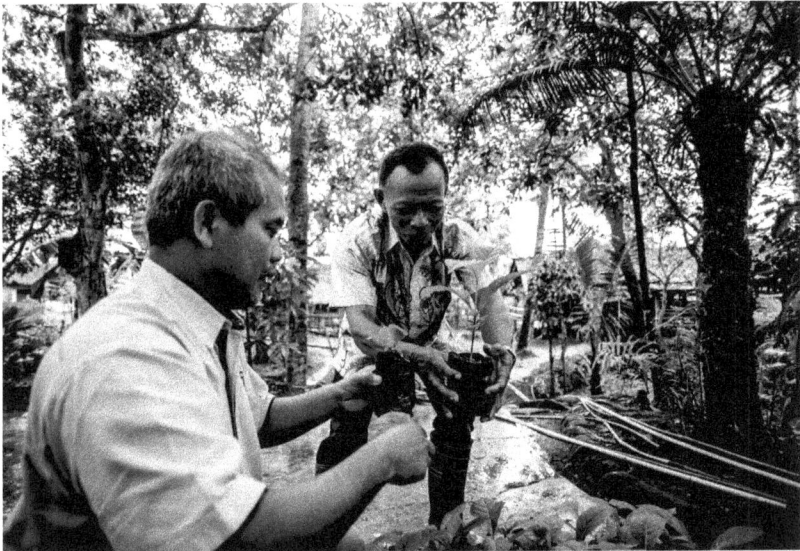

*Seedlings for jungle planting, Central Java, Indonesia: stepping up tree planting to enhance carbon storage a valuable strategy*

---

226 For further information go the Intergovernmental Panel on Climate Change website: http://www.ipcc.ch/

Turning now to the critical issue of sustainable jungle management, in earlier chapters we have seen that despite much deliberation and a wide range of supporting initiatives, implementing sustainable management continues to be elusive. We have also seen that there are underlying, complex issues that make the goal of more sensitive and sustainable practices difficult. We have considered issues like undefined or overlapping property rights, weak governance and traditional demand for timber that have led to high levels of insensitive and unsustainable logging.

Across the temperate forest world many countries have revised their forest policies and operations to incorporate sustainable forest management practices. While there are pleasing trends in this direction, in the tropics a wide gap still persists between what is visualised in policies and what actually happens in practice.

We have also recognised that weak governance and corruption continue as a major challenge, especially where political and institutional frameworks remain fragile. For example, while since about the year 2000 the extent of legally protected jungle across Southeast Asia has remained reasonably stable, the management of these areas continues to be problematic because of ongoing encroachment, illegal clearing, poaching of animals and illicit logging.

We have acknowledged the overarching need to strengthen governance and to reduce corruption, bribery and extortion. Better governance will increasingly become a prerequisite for continuing to export to consumer country markets and for wood processing investment. Countries with poor governance will be severely disadvantaged in competing for funds, with money gravitating to countries where investors have confidence in government institutions and their performance.

While the science is becoming increasingly advanced in terms of the role of Southeast Asian jungles in climate change abatement, additional research and development effort related to land use planning, sustained yield timber management and wood product innovation is needed. There is little doubt that further developments in science and technology could significantly improve jungle survival prospects and sustainable forestry activity.

Technologies like tree breeding and the development of new products and processes are key requirements to lift forest productivity and the commercial success of wood processing. Remote sensing methodologies are now also able to revolutionise the ability to monitor jungles and help track changes on a real time basis.

So future sustainable management points to the need to enhance collaboration between science, policy, industry, conservation and other areas of effort. Stronger science inputs together with a focus on translating existing knowledge into technologies that are practical, understandable and able to be implemented will be part of the equation. It will also be vital that future land use planning better recognises the values of ecosystem services, like carbon sequestration, water provision and sources of livelihood for rural communities.

As we saw in the last chapter, economic growth across Southeast Asia has led to reductions in poverty. Despite this growth, poverty is likely to continue to be an issue in some countries. Also in some countries economic growth has actually exacerbated disparities between rural and urban areas.

As we draw this book to a close we should emphasise the essential role tropical jungle sourced non wood products have and continue to play in the economic and social well being of many people, especially tribal communities and the rural poor. Non wood products make a major contribution to the subsistence needs of jungle dependent communities and to poverty alleviation. Improvements in processing technologies have led to the domestication of a number of products and to income opportunities. Further, the rights of indigenous people to manage their ancestral jungle homelands is now increasingly being recognised and practiced.[227]

Past experience and present reality is that land use planning decisions in Southeast Asia can be driven by vested interests directed at immediate economic gains. It is expected that forest degradation will remain a major problem, especially in more densely populated, low income countries where dependence on land and forests is high. Future land use planning decisions need to take into consideration longer term social equity, environmental sustainability and the recognition of the need to maintain the natural capital of the jungles – balancing shorter term economic goals against environmental and social sustainability. Easy to say – much more challenging to achieve.

---

227 *The Future of Forests in Asia and the Pacific: Outlook for 2010*, 2009, Food and Agriculture Organization of the United Nations Regional Office for Asia and the Pacific, Bangkok, Thailand

*Inspecting sustained yield management, Malaysia: recognition of the need to maintain the national capital of jungles*

Recognising that populations will continue to grow and levels of consumption increase, it is imperative that Southeast Asia invests in conserving and enhancing the management of its natural asset base. While the region is unlikely to face timber product shortages in the immediate future, rebuilding the natural resource base of jungles and conserving these resources must be a high priority.

Efforts to achieve consensus about how jungles should be managed and for what purposes will be a key element in regional land use planning over coming decades. Greater efforts will be required to integrate public opinion into decision making, and to build levels of awareness in relation to jungles and forestry so that policies can be implemented with broad community and government support. This effort should, for instance, see the removal of some present bizarre incentives that encourage land conversion and underprice exports of jungle logs and semi processed wood products.

Without concerted effort further over exploitation and continued degradation of the jungle estate appears likely, as is the loss of irreplaceable biodiversity. Undoubtedly the stewardship of the spectacular jungles of Southeast Asia and the irresponsible and reckless loss of iconic regional animals by the current generation will be judged harshly by our successors.

*Threatened tiger, Sumatra: without concerted effort irreplaceable biodiversity loss likely*

Carefully targeted investments will be a key ingredient of future jungle survival and improved management. Such investments could generate a significant number of new jobs and income in forestry and wood processing industries. Much of this employment would most likely be in rural areas, where it would impact positively on rural poverty. Employment in sustainable management activities has a double benefit of building the natural asset base and reducing the deforestation that occurs when other income earning opportunities are absent.

As this chapter comes to a close I am conscious that it is easy to be gloomy, pessimistic, despondent and even desperate about the future prospects for the jungles of Southeast Asia. However, in writing this book I want to be positive – to be optimistic – otherwise I wouldn't have bothered.

So a combination of the actions we have traversed, plus more judicious government across Southeast Asia is the way to a better future. I do not believe with what the jungle estate brings to the region's economic, social and heritage, the people of Southeast Asia will not let their precious jungle resource and the economic corner stone it provides, slip through their fingers. Don't you let that happen either!

So let us be crystal clear in case there is any doubt, making tropical jungle trees too valuable to destroy is a big piece of the solution to their survival, along with continuing to prosecute the case for environmental services to gain a harder commercial 'edge'. Making sustainable management and value adding timber manufacturing score well in economic terms will also help ensure that jungles remain intact.

*Teak chairs, Yangon, Myanmar: making sustainable management and value added timber manufacturing score well in economic terms*

As Amha Bin Buang, Assistant Director of the International Tropical Timber Organization said in Jakarta in September 2013[228] attempting to terminate the trade in tropical timber as the answer to the restoration of tropical forests is not a solution. He expressed the view that discontinuing trade will in all likelihood lead to an acceleration of clearance and degradation because other forms of land use will, by default, become immediately much more attractive economically.

He suggested that part of the way forward was:

> ... to soldier on and persevere in restoring and strengthening the international trade in tropical timber. This entails addressing urgently and effectively the acute image problem which tropical timber has been suffering.

In making these comments he placed priority on tropical timber producing countries and the international timber trade adopting a positive and proactive approach to addressing the challenges of weak forest governance, and on more effectively reining in illegal logging and related trade.

---

228 *What is happening to the tropical timber trade?* 2013 Amha Bin Buang, International Tropical Timber Organization. Address to the 3rd High Level Market Dialogue: The new era of Indonesian legal timber products to meet global markets, Jakarta, Indonesia.

# EPILOGUE

## AS IT WAS IN THE BEGINNING

Finally at the end as at the beginning – or more correctly from the Bible's book of Matthew – as it was in the beginning, so shall it be in the end. We are now at the conclusion of this story and should close the loop by reflecting back to the beginning, and on the voyage we have taken through the jungles of Southeast Asia. I hope by now the task ahead for us is clear and compelling.

The issues and challenges connected with sustaining the jungles of Southeast Asia that we have traversed might seem to be enormous and insurmountable, and so they might be. However, each step forward represents one step less on a difficult, but essential journey. Sure, governments and communities across Southeast Asia have a clear and compelling role in confronting obstacles and implementing the solutions we have outlined, but we also have a job to do. By our buying and lifestyle choices, advocacy and concerns we can add our voices to those of others. Yes, you can, for the need to assist in sustaining the jungles of Southeast Asia is urgent and critical. You can speak for the iconic animals and tall tree ecosystems we have discussed – and for the rural poor and ancient tribal communities.

I said in the Prologue – do you remember – that this book has an important mission. I put this as simply as I could – the need to better recognise the incredible values, worth and potential of Southeast Asian's tropical jungles, and to figure out what we might do to ensure that these values were recognized, quantified and sustained. We now know that to stand any prospect of long term survival the plants and animals calling these jungles home must be sensitively managed and carefully perpetuated.

The proposition that the destruction of tropical jungles has in part been driven by grinding poverty and the legitimate wish of Southeast Asian countries to improve their economic and welfare standards has been well established. It is about survival or money or both. We have also been able to confirm, in blunt terms, that the motivating factor behind ongoing jungle destruction has and continues to be the hard economic reality that tropical jungles have little inherent economic value.

*Tropical jungle: must be sensitively managed and carefully perpetuated*

So despite descriptions of jungles being precious, priceless and irreplaceable the economic logic has been to convert them into something more economically 'useful', or otherwise, just accept that they will be whittled away by the increasing millions of rural poor. Destruction continues to be motivated by the reality that in conventional economic terms jungles have little tangible value and should therefore be converted to some other use that is capable of earning a dollar.

We have been able to identify and discuss the economic options that might save jungles from continuing devastation and decline. Fundamentally it is apparent that there must be strong economic incentives to retain and perpetuate jungles rather than destroy them. We have considered both traditional and emerging economic options that may offer part of a solution to retaining jungles from the steady advance of palm oil plantations; conversion to other forms of agriculture or piece meal destruction.

We now have a sound basis to argue for long term safeguards and for managing tropical jungles that is founded on an economic case absent from past debates. We have established that in relation to promoting the cause of jungle survival economic opportunities lie in the increasing prospects of sustainable, legally verified timber production and in carbon credit trading. These activities provide an opportunity for assigning measurable, real dollar values to jungles and provide the capacity to disperse income generated across governments, companies and communities.

*Wooden table and chair: value adding products part of the economic case*

In addition to more conventional outcomes this economic case could yield innovative, value added wood based products; save the lives of those who share the planet with us, and achieve some critical climate change benefits. So putting it in simple terms, it is about establishing and implementing the case for developing strong economic incentives to retain healthy, functioning economically viable jungles across Southeast Asia. This economic scenario will help to create a set of circumstances where jungles are seen as an economic asset, not a liability, and where governments, corporations and communities have a vested interest in their conservation.

As suggested in the last chapter, it is easy to be gloomy, pessimistic, despondent and even desperate about future prospects for the jungles of Southeast Asia. But we know better than that. We know there is a need to be positive – to be optimistic – so a combination of the actions we have traversed through preceding chapters offer the prospect of a better tomorrow for the jungles of today.

At this concluding stage in proceedings let us be crystal clear – making trees too valuable to destroy is the critical piece of the solution to jungle survival. Making sustainable management and value adding timber manufacturing score well on the economic scale will help ensure that jungles remain. Continuing to prosecute the case for environmental services focused on gaining additional commercial benefits will also assist in delivering compelling environmental outcomes.

Finally, thanks for sharing this journey through the jungles of Southeast Asia with me. Call me boring, but I don't mind being called "the tree man" by some of my friends. As best I can I attempt to talk for trees and forests – we are on this planet by their grace and favour – we have a moral obligation and an economic imperative to treat them with respect, humility and wonder. They can do without us, but we cannot do without them.

# ACKNOWLEDGEMENTS

Without reservation in writing this book I acknowledge I have climbed on the shoulders of others. They include early disciples of sustainable forest management, researchers, foresters, conservationists and more recent advocates of sensitive jungle management. They have all contributed to a recognition of the need for jungles to have economic worth so that trees can be kept standing and vital habitats maintained.

I have attempted to interpret the writings of others and to listen and learn from my working visits to jungles across Southeast Asia. I am grateful for the expertise, experience and enthusiasm of those with whom I have interacted. Can jungles continue to make a valuable contribution to the welfare of humankind and at the same time retain their majesty, diversity and wonder? That is the question – that is the challenge.

I should also thank others who have helped to get this book into your hands. Notably Jan Hume provided great assistance with fine tuning the manuscript so it passed muster with the publisher. Russell Jeffery of Emigraph Creative in Sydney provided welcomed help in preparing the illustrations and sourcing some images.

Critically Anthony Cappello and the team at Connor Court Publishing kept the faith and were responsible in having this book designed, published and marketed in both printed and electronic formats.

My grateful appreciation to you all.

# Picture Acknowledgements

**Picture credits**

Every endeavour has been made to identify and attribute the source of pictures. The source of some pictures obtained from the internet has in some cases not been able to be established and acknowledged:

*Page Numbers:*

Anon 9, 21, 40, 47, 52, 64, 65, 66, 109, 124, 146, 147, 156, 162, 164, 190, 191, 226

Ensis 208

Forest Research Institute Malaysia 170

Forestry Tasmania 24, 48

Hardi Baktiantoro 7, 86, 117, 148, 150

John Halkett 8, 12, 13, 16, 20, 23, 27, 36, 39, 41, 166, 115, 120, 121, 122, 160, 166, 174, 180, 186, 187, 197, 210, 223, 227

J.L. Kendrick 37

Justin Mott (courtesy The Nature Conservancy) 82, 87, 216

Michael Mullan 12

NASA 10

Shutterstock 28, 30, 31, 42, 43, 45, 50, 51, 62, 64, 75, 85, 96, 111, 116, 117, 127, 130, 154, 198, 201, 222

Stora Enso Timber Australia 206

US Department of Defense 134

Western Red Cedar Export Council 189

World Agroforestry Centre 218

Illustrations by Russell Jeffery, Emigraph Creative, Sydney.

# INDEX

www.ingramcontent.com/pod-product-compliance
Lightning Source LLC
Chambersburg PA
CBHW061013280326
41935CB00009B/951